Applied Reliability for Industry 1

Reliability of Multiphysical Systems Set

coordinated by
Abdelkhalak El Hami

Volume 16

Applied Reliability for Industry 1

Predictive Reliability for the Automobile, Aeronautics, Defense, Medical, Marine and Space Industries

Edited by

Abdelkhalak El Hami
David Delaux
Henri Grzeskowiak

iSTE

WILEY

First published 2023 in Great Britain and the United States by ISTE Ltd and John Wiley & Sons, Inc.

ISTE Ltd
27-37 St George's Road
London SW19 4EU
UK

www.iste.co.uk

John Wiley & Sons, Inc.
111 River Street
Hoboken, NJ 07030
USA

www.wiley.com

Library of Congress Control Number: 2022944183

British Library Cataloguing-in-Publication Data
A CIP record for this book is available from the British Library
ISBN 978-1-78630-691-3

Contents

Foreword

Predicting and then guaranteeing the reliability of an industrial system is a major challenge for manufacturers in the automotive, aeronautics and defense industries, as well as for those in the rail, telecommunications, nuclear and medical sectors, among many others. But above all, it is a major challenge for us, who use this equipment on a day to day basis, who must have absolute confidence in the information transmitted and the decisions made in real time. The increasing development of connected objects (autonomous cars, home automation, etc.) will lead to a drastic reduction of human intervention in favor of the intervention of mechatronic systems. All these systems can only be deployed if users have absolute confidence in the reliability of the equipment.

These forms of equipment are divided into two main functionalities: a hardware part mainly composed of electronic boards (coupled with mechanical systems) and a real time software part enabling the implementation of the equipment and the realization of the expected tasks.

Predicting and guaranteeing the reliability of electronic equipment is a huge and endless task. On the one hand, the number of component types used to build these boards is very high and on the other hand, the new functionalities in innovative equipment require numerous reliability and robustness tests.

Thanks to the support of the Astech and MOV'EO competitiveness clusters, the NAE aeronautics industry, the Normandy and Ile de France regions, and the Rouen and Versailles Chambers of Commerce, programs have been implemented and have produced exceptional results (AUDACE, FIRST-MFP, PISTIS, CRIOS, SYMSTECK, etc.).

Given the richness of the results of these three books that have been published, I would like to thank very warmly Abdelkhalak El Hami, David Delaux and Henri Grzeskowiak for their remarkable work in the implementation of six volumes in total as well as all the authors who have devoted many hours to shaping all their results. This has meant that this essential information does not remain the property of few but shared by the largest number of engineers, technicians, researchers and students.

The first volume is devoted to Predictive Reliability, for estimating or predicting the reliability performance of systems.

The second volume is dedicated to Experimental Reliability using tools and test methods to demonstrate the reliability of systems.

The third volume presents Operational Reliability; it verifies the reliability performance of systems in real life by way of an analysis of the field data.

I would like to thank all those involved in this program as well as the financiers (State, Regions) without whom these projects could not have succeeded and I wish that the members of the hard core set up in these programs (AUDACE, FIRST-MFP, PISTIS, CRIOS, SYMSTECK) continue their association because their skills will be essential for guaranteeing the reliability of the very many technologies which are in the process of emerging and which will equip the mechatronics systems of tomorrow.

Philippe EUDELINE
President of Normandie AeroEspace (NAE)
October 2022

Preface

March 2020 was a turning point in our lives and for industries worldwide when Covid-19 struck the world! This pandemic has highlighted the *fragility* of our industries while exacerbating the paradigms of development, innovation, commercial competition and the acceleration of research phases.

More than ever, we need to be *agile* in our American, European, Asian and even global business plans, while at the same time *cementing* the use of our products in their operational use. It is clear that the *challenges* of combining the concepts of variability and stability, standardization and acceleration of product development are enormous. The "post-Covid" world is pushing all industries to the very limits of reliability engineering.

It is true that the word "reliability" is often used as a marketing cosmetic without any real care for its underlying scientific substantiation! Nonetheless, behind this word lies an applied science, validated by analysis, research and talented people.

Which type of the reliability approach can a designer, an engineer and a manager use in their professional environment? Whatever the industrial field (aeronautics, space, defense, automotive, home automation, etc.), how is it possible to design a reliable product more quickly, all the while upholding safety measures?

This book illustrates, in great detail, the multidisciplinary, state-of-the-art science of predictive reliability. Many experts are convinced that reliability is not limited to the statistical sciences. Various disciplines interact in order to bring a product to its highest possible level of reliability, as available through today's technologies, developments and production methods.

In the context of perpetual research, in order to improve industrial competitiveness, the evolution of product design methods and tools appears to be a strategic necessity with regard to the need to reduce costs. Nevertheless, the reduction of design costs must not be done at the expense of reliability for the new systems being proposed, which must also make significant advancements.

To increase the competitiveness of their mechatronic devices, automotive and aerospace suppliers need to find innovations, for both the design and assembly processes, in order to reduce product development times. In addition, these innovative products must combine excellent functional and operational performance, which includes reliability, to fully meet the expectations of the global market. For car manufacturers, these reliability expectations will certainly increase with the intensified market penetration of electric or hybrid vehicles by 2020–2025, estimated at 10% of the market. In addition to these expectations for operational reliability, there is also the need to quickly eliminate the risks of immaturity associated with product innovations. This need is strongly linked to reducing vehicle development times to a strict minimum.

In the field of aeronautics, the needs are mainly related to the forecasting and control of costs induced by failures occurring at the commissioning phase, during the warranty period and during the operation of the aircraft, with the knowledge that, in the future, new aeronautics equipment sales contracts will be increasingly oriented toward *sales by the operating hour*. Although the aeronautical sector has relatively low production volumes when compared to the automotive industry (in terms of the number of units per type of product), the financial stakes are high and, as a result, aeronautical manufacturers are over-sizing mechatronic components to counteract problems they are currently unaware of. Reliability is defined by the international standard ISO/TR 12489:2013[1] as: "The probability for an item to perform a required function under given conditions over a given time interval".

This book aims at proposing new methods to analyze the lifecycle of a system. The lifecycle enables us to record the development phases according to the development time and the levels of complexity for its integration.

The three main families of activities governing the development of reliability for a mechatronic system are predictive reliability, experimental reliability and operational reliability.

Predictive reliability includes all engineering activities used to *estimate or predict* the reliability performance of the final mechatronic system. Predictive reliability is particularly focused on in *Applied Reliability for Industry 1*.

---

1 [ISO 13] ISO/TR 12489:2013, Petroleum, petrochemical and natural gas industries – Reliability modelling and calculation of safety systems, ISO/TC 67, edition 1, 2013.

Experimental reliability includes all the tools and testing methods used to *demonstrate* the reliability of the mechatronic system. Experimental reliability is advanced in the *Applied Reliability for Industry 2*.

Operational reliability *verifies* the reliability performance of the mechatronic system in its real life through an analysis of the field data. Operational reliability is presented in the *Applied Reliability for Industry 3*.

The first chapter in *Applied Reliability for Industry 1* presents FIDES, a method for evaluating and constructing the reliability of electronic systems. The second chapter presents "Reliability in Maritime Transport: Choosing a Container Handling System". Chapter 3 presents the "Generation of a Failure Model Through Probabilistic 'Stress–Strength' Interaction in a Context of Poor Information". Chapter 4 is devoted to "Reliable Optimization of Dental Implants Using the Generalized Polynomial Chaos Method". Chapter 5 presents "Multi-Objective Reliability Optimization Based on Substitution Models. Applied Case Study of a Hip Prosthesis". Chapter 6 proposes an adaptation of the CMA-ES method that is assisted by the *kriging metamodel* for the global optimization of a problem. The implementation of this proposed algorithm has been carried out for the global optimization of solder joints for a PQFP housing. The results of numerical studies show that the proposed KA-CMA-ES algorithm is more efficient and effective than the standard CMA-ES algorithm. Chapter 7 presents the "Reliable Optimization of Vibro-Acoustic Problems in the Presence of Uncertainties via Polynomial Chaos".

To conclude, there is a *before* and an *after* Covid. The reliability challenges for our industries to address in this new period are significant. The global context but also the green, sustainable and carbon-neutral goals are the next steps in a fierce economic competition that will span the next decade. The authors hope that *Applied Reliability for Industry 1* on predictive reliability will impart meaningful reflections and help to guide readers toward the best practices to encourage efficient design and effective decision-making.

This book is dedicated to all those who have lost a loved one during the Covid-19 pandemic worldwide.

Abdelkhalak EL HAMI
David DELAUX
Henri GRZESKOWIAK
October 2022

1

FIDES, a Method for Assessing and Building the Reliability of Electronic Systems

Reliability has become a key performance of current developments. It is now, more often than not, a contractual performance in its own right, generally associated with strong commitments, monitored on-the-ground and subject to severe penalties. Apart from these contractual aspects, reliability remains a crucial parameter for the overall cost of ownership and availability of a product.

On the other hand, predictive reliability is an essential input for most quantified operational safety studies, such as the sizing of the support system: FMEA/FMECA (failure mode, effects and criticality analysis), probability of correct functionality, maintainability, testability and availability.

Lack of confidence in predictive reliability taints all analyses that depend on it with uncertainty. This can have profound negative consequences on product development (e.g. by leading to the erroneous rejection of one technology in favor of another, or by dictating inappropriate architecture choices), but also on costs. It is therefore imperative to have realistic predictive reliability assessments. The FIDES reliability assessment and construction method [FID 09] (from the Latin "confidence") meets these requirements for electronic systems.

Chapter written by Franck DAVENEL.

For a color version of all the figures in this chapter, see: http://www.iste.co.uk/elhami/applied1.zip.

1.1. The inadequacy of existing methods

Since 1965, MIL-HDBK-217 [MIL 95] has been one of the historical standards long used by the vast majority of military equipment manufacturers (but not only by these) to predict the reliability of electronic equipment. Today, this method has become inadequate as a result of the evolution of components and the usage of systems, forcing many users to implement adjustments (more or less precise and justified) in order to make the prediction as accurate as possible.

In addition to the FIDES methodology, there are three main predictive reliability methods that are commonly used.

1.1.1. *MIL-HDBK-217F [MIL 95]*

This method, often presented as a standard but which in fact is only a handbook, still remains the flagship in the field. However, it is largely obsolete. It has not been maintained since 1995 and is generally based on user experience from the 1980s. Its content is no longer representative of current technologies. The MIL-HDBK-217 guide has been adapted to certain specific fields, such as telecommunications in the United States (Bellcore/Telcordia SR-332).

1.1.2. *UTE-C-80810 (or RDF2000, or IEC 62380 TR Ed.1)*

This compendium of reliability covers most of the recent technologies [UTE 00]. Nevertheless, it suffers from certain shortcomings that do not allow it to provide an all-inclusive answer. In particular, RDF 2000 is inadequate for dealing with non-benign environments. It does not enable us to distinguish the different suppliers of the same item, nor the level of manufacturer processes. The IEC 62380 standard, which is not maintained, was recently withdrawn from the international IEC standard.

1.1.3. *PRISM® or 217plus*

This is a commercial software product, initially developed by RAC, of which another version (217plus) is now being marketed by QUANTERION [DEN 97]. This tool proposes an approach to predictive reliability that breaks tradition with MIL-HDBK-217. However, it has several drawbacks that prevent its recognition. At the end of the day, its use is limited outside the United States. The proposed calculation models are rather poor. The coverage of recent component families is still quite limited, and the availability of a database of raw failure rates is only a meager counterpart. The 217plus method does not allow us to deal with complex life profiles (at most one operating phase) or to distinguish between different suppliers of the same item.

The existing methods are therefore inadequate to meet the needs expressed above. Moreover, none of these methods can really be used as support for a reliability engineering activity. Finally, we can see in Figure 1.1 that only FIDES has evolved in the last few years with a new version under construction and with a possible release in 2023.

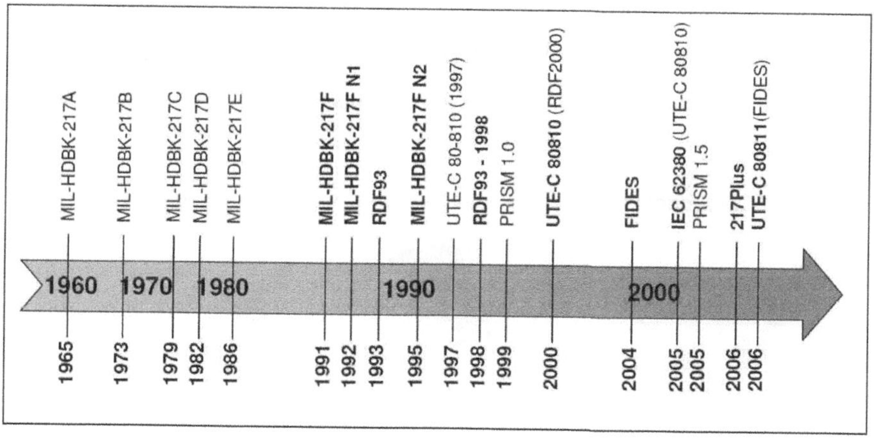

Figure 1.1. *History of electronic predictive reliability collections*

1.2. The ambition of FIDES

In order to have a new reference system, the DGA (*Direction Générale de l'Armement:* "Directorate General of Armaments") brought together a consortium of eight major aeronautical and defense companies in early 2000: AIRBUS, Eurocopter, MBDA Missiles Systèmes, Nexter Systems, Thales Research and Technology, Thales Airborne Systems, Thales Avionics and Thales Underwater Systems.

The first objective of this study was to have a new reliability assessment method for electronic components that is realistic and takes into account new technologies. The second objective was to provide a reliability engineering guide to support manufacturers in the development of new electronic systems.

The main objective of FIDES is therefore to have a reliability assessment method for electronic components that is an alternative to the MIL-HDBK-217 standard and that takes into account new technologies, in particular, the consideration of

Commercial Off-The-Shelf (COTS) components that are increasingly used. One of the characteristics of FIDES is the identification and consideration of all technological and physical factors that have an impact on reliability. This guide also considers failures related to development, manufacturing, operation and maintenance processes, as well as support activities during the entire product lifecycle. It is consolidated by analyses of test data, user experience and existing models.

In this context, the purpose of the FIDES guide is to propose a new methodology that is:

– realistic and accurate;

– usable for both the estimation and construction of reliability;

– usable for any type of item (components, assemblies, sub-assemblies), including COTS, as illustrated in Figure 1.2.

Figure 1.2. *Illustration of the electronic items covered by the FIDES guide*

A first version of the guide was published in 2004 (FIDES 2004A) and an update was published in 2010 (FIDES 2009A) [FID 09]. The FIDES guide is available free of charge on the FIDES website (www.fides-reliability.org) and is well distributed in the French industry and throughout the rest of the world as well.

Since its publication, FIDES has aroused great interest. As a result, the *Directorate General of Armaments* has approved the method as a "standard applicable to French military programs"; the guide has been integrated into the *Référentiel normatif des programmes d'armement français* (the "normative reference framework for French armament programs") and, since 2011, into the European Defense Agency's (EDSTAR) reference system. In addition, the board of directors of the UTE (now AFNOR) accepted the FIDES guide as a technical document at the end of 2004, under the reference UTEC 80-811. In France, FIDES is therefore accessible to anyone as a recognized organization in the field of standardization, which allows for a wider distribution to organizations outside of the consortium (equipment manufacturers, CNES [VER 05a, VER 05b]). The extension to an international document is underway within the IEC through a group of international experts representing 22 nations. The international FIDES standard under the reference IEC 63142 was officially released in December 2021. In addition, we have noted the interest shown by many international companies, notably American companies such as Boeing [BEC 04] or Raytheon, but also major automotive manufacturers in Europe. A tangible sign for the utility of FIDES is that most of the publishers for operational safety software throughout the world now offer a FIDES module.

1.3. General presentation of the FIDES method

The main features of FIDES are as follows:

– models for EEE components (electrical, electronic, electromechanical) as well as for electronic boards or some sub-assemblies, and for recent technologies;

– a methodology applicable to both military or space components (so-called high reliability) and COTS and for all fields (military, industrial or civil);

– the identification and consideration of all technological and physical factors that have an identified role in reliability;

– models based on the physics of failures, which are not only based on experiential feedback but also on the exploitation of reliability tests (active) or manufacturer data (sub-assemblies);

– precise consideration of the life profile, as opposed to the use of standard predefined environments;

– consideration of accidental electrical, mechanical and thermal overloads (or overstress);

– taking into account failures related to the development, production, operation and maintenance processes;

– the possibility of distinguishing several suppliers of the same component;

– easier updating, and the possibility to deal with new technologies without waiting long for user feedback.

The FIDES reliability approach described in the publication [GUI 04] is based on the consideration of the three components (technology, process and usage) as represented in Figure 1.3, which represents the "*logo*" of the method. These components are considered throughout the entire lifecycle, from the product specification phase and through the operation and maintenance phases. "Technology" covers both the technology of the item itself and the equipment into which it is integrated. The "Process" considers all the practices and rules of the art, from the specification of the product to its replacement. Finally, "Usage" takes into account both the constraints of use defined by the design of the equipment manufacturer and those in operation by the end user.

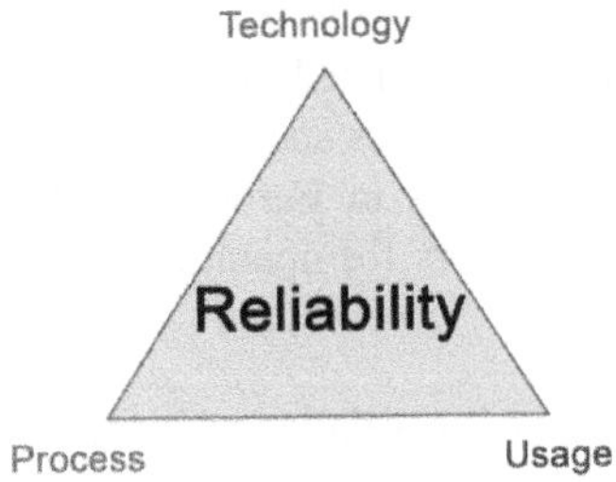

Figure 1.3. *Illustration of the FIDES methodology with its three contributors*

The proposed models therefore consider a "Technology" subject to "Usage" constraints according to an approach that takes into account the failure mechanisms and associated contributors, and above all, they weigh the risk of failure by all the "Process" contributors that can activate, accelerate or reduce the effects of these mechanisms.

By identifying the contributors to reliability, whether technological, physical or process-based, the FIDES methodology is able to act according to definitions throughout the whole lifecycle of the products so as to improve and control the reliability.

1.3.1. *Failure rate*

The FIDES evaluation model assumes that the failure rate is constant, at least during the so-called service life phase. The periods of youth and old age are excluded from the forecast. Strictly speaking, the failure mechanisms do not respond to a modeling of their likely occurrence through a law whose parameter (failure rate) would be constant. However:

– many failure mechanisms, although cumulative, which therefore increase over time, have such a dispersion in their distribution that they can be assimilated to a constant over the periods being considered;

– the multiplicity and diversity of the components, even on a single Printed Circuit Board Assembly (PCBA), will lead to an accumulation that is very close to a constant;

– the differences in the lifecycle situation of the components between the equipment of a same system, or a multiple system, result in a constant rate for the observer of the system level.

For these reasons, the use of a constant failure rate remains the most relevant approach for estimating the predictive reliability of electronic systems.

The inadequacy of available predictive reliability methods, and in particular MIL-HDBK-217, has often led to the use of a constant failure rate being considered fundamentally inadequate, instead favoring deterministic life prediction approaches (e.g., the simulations proposed by CALCE, University of Maryland, USA [BAU 94]).

In reality, these two approaches are essentially complementary.

1.3.2. *The structure of FIDES models*

The general FIDES model allows for the calculation of failure rates for electronic equipment before any consideration of redundancy or architecture.

The overall failure rate of the electronic equipment is obtained by adding up the failure rates of each of its components, thus considering a serial system. We have the usual relationship of:

$$\lambda_{\text{product}} = \left(\sum_{\text{Item}} \lambda_{\text{Item}} \right) \qquad [1.1]$$

The FIDES methodology is based on four key components:

– the physical failure rate, which reflects the contribution of technologies and constraints: $\lambda_{\text{Physical}}$;

– induced failures (accidental overloads): Π_{Induced};

– the manufacturing quality of the part: $\Pi_{\text{Part\_Manufacturing}}$;

– the level of control and construction of reliability in the lifecycle: Π_{Process}.

The general equation for FIDES models is given as:

$$\lambda = \lambda_{\text{Physical}} \cdot \prod_{\text{Part\_Manufacturing}} \cdot \prod_{\text{Process}} \qquad [1.2]$$

with:

$$\lambda_{\text{Physical}} = \left[\sum_{\text{Physical Contributions}} \left(\lambda_0 \cdot \prod_{\text{acceleration}} \right) \right] \cdot \prod_{\text{induced}} \qquad [1.3]$$

where

– λ_0 is the basic failure rate, and also representative of the characteristics of the technology;

– $\Pi_{\_acceleration}$ is an acceleration factor reflecting the sensitivity to a physical contributor: electrical stress, temperature, thermal cycling, humidity, mechanical, chemical environment (specified nominal contributions);

– $\Pi_{Induced}$ represents the contribution of induced factors (also called accidental overloads or overstress), typically expected in a given application (EOS, ESD, etc.).

The details and justification of the construction of the models are described in [BAY 10].

1.3.3. *Physical models*

The models used in FIDES to translate the failure rates of the different parts are based on the physics of failures and supported by the analyses of test data, user experience and existing models. The FIDES methodology is thus different from classical methods (MIL-HDBK-217), which were predominantly developed from the statistical exploitation of experiential feedback.

Therefore, the models proposed in FIDES are representative of current technologies and processes, and not those of equipment which have already been in service for a long time. On the other hand, the exploitation of the physics of failures allows FIDES models to explicitly describe the factors that affect reliability.

Two methods were used to build the models for the components covered by the methodology.

1.3.4. *The exploitation of manufacturer data*

This method has mainly been implemented among active circuits. It exploits reliability test results published by manufacturers for thermal, humidity and thermal cycling aspects of packages. It uses qualification tests and simulation results of package reports (results from partners and manufacturers) for the thermal cycling of solder joints.

Table 1.1 shows the types of tests that were used to quantify the contributions of thermal stress, humidity and thermal cycling.

Acronym	Signification	C.I.	Discrete	λ_0
HTOL	High temperature operating life	X		Thermal
HTRB	High temperature reverse bias		x	Thermal
HTGB	High temperature gate bias		x	Thermal
PCT or AC	Pressure cooker test/autoclave	X		Humidity
TC/TS	Temperature cycle/ thermal shock	X	x	Cycling Th. housing

Table 1.1. *Reference tests used to define baseline failure rates*

1.3.5. *Exploiting databases of failure mechanisms (not failure rates)*

This methodology has been used for components other than assets. It allows for the compilation of contributions to physical constraints related to the different mechanisms. Figure 1.4 presents the case of the gelled tantalum capacitor.

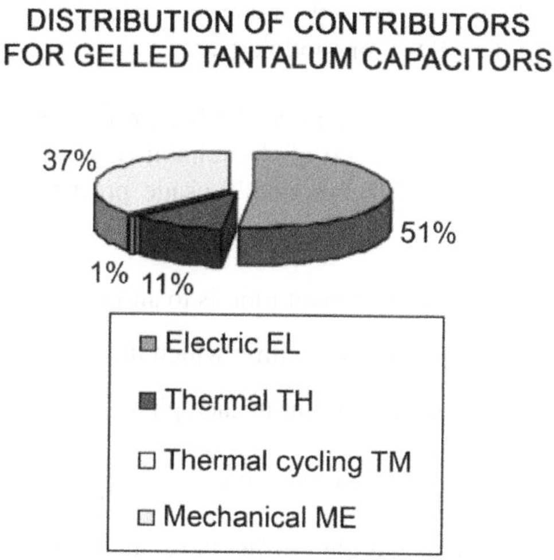

Figure 1.4. *Distribution of contributors for gelled tantalum capacitors*

The final formatting of the models from these distributions ultimately requires:

– assigning a base failure rate (from existing collections and/or experiential feedback);

– identification of technological contributors (e.g. the type of case used for gelled tantalum capacitors) that may influence reliability.

In all cases, the modeling of environmental effects ($\Pi_{\_acceleration}$ factors) was based on classical acceleration laws: Arrhenius (thermal stress), modified Peck (humidity stress), modified Coffin–Manson, Engelmeyer and Norriz–Landsberg (thermal cycling stress) and modified IPC-SM785 (mechanical stress).

The models for assemblies and other sub-assemblies are built according to specific approaches which respect the general principles listed above. These models therefore have exactly the same structure as the component models.

These physical models allow predictive reliability assessments to be made for both severe and non-operating or storage environments.

1.3.6. *Life profile*

The detailed and explicit consideration of physical constraints and their relative duration, as opposed to the use of predefined standard environments, is a fundamental feature of the FIDES methodology.

This requires us to precisely determine the life profile for which the reliability calculation is desirable. The accuracy of the reliability assessment depends on the accuracy for the description of the product's usage profile within an operational system.

The principle of the life profile constitution is to simply:

1) divide the life profile into phases within which the constraints are constant;

2) quantify the level of each constraint in each phase.

Table 1.2 shows an example of a life profile.

Given this consideration of the life profile, the calculated failure rates are ideally expressed per calendar hour (rather than per operating hour for example). This unit has the merit of universality. However, the FIDES methodology also enables us to calculate the failure rate on an isolated phase of a usage profile, or to convert the calendar failure rate according to another unit (/operating hours, /flight hours, /kilometers, etc.).

		Thermal and humidity		**Thermal cycling**						**Mechanics**
Phase title	Calendar time (hours)	On/ off	Ambient temperature (°C)	Humidity level (%)	ΔT (°C)	Number of cycles (/an)	Cycle time (hours)	Maximum temperature during cycle (°C)	Random vibrations (Grms)
① Ground-operation ON/OFF	700	On	40	30	25	350	2.00	40	0.05
② Ground-operation scale	1,400	On	55	30	15	700	2.00	55	0.05
③ Ground-rolling	630	On	40	10	–	2,100	0.30	–	5
④ Flight-ascent/ descent	1,050	On	40	10	–	1,050	1.00	–	0.6
⑤ Flight-stabilized	3,150	On	40	10	–	1,050	3.00	–	0.6
⑥ Ground-dormant	1,830	Off	15	70	10	365	5.01	20	–

Table 1.2. *Example of a life profile for a bay computer of a medium-haul civil aircraft, from the FIDES guide*

1.3.7. *Other contributors*

1.3.7.1. *Induced failures (Π_{Induce})*

Induced failures are failures caused by accidental overloads (or *overstress*).

An accidental overload is an event that is outside the normal scope of the product's usage.

The induced factor (Π_{Induce}) is based on four subfactors:

1) the sensitivity of the technology to overstress;

2) the placement of the part within the application (specifically: interface or not);

3) the type of application, which defines the exposure to overstress;

4) the strengthening efforts made in development (robustness construction).

The FIDES approach takes into consideration the possibility of accidental overloads of a thermal, mechanical or electrical nature.

1.3.7.2. *Quality of part manufacture* ($\Pi_{part\_manufacturing}$)

Taking into account the manufacturing quality of the item (component, board, sub-assembly) has two main objectives:

1) interpret the impact that the quality and technical control employed in the manufacturing of the part has on reliability;

2) discriminate between manufacturers.

1.3.7.3. *Reliability process* ($\Pi_{process}$)

Operational observation clearly shows a preponderance of failures attributable to the various lifecycle activities when compared to failures that are completely attributable to the components (see Figures 1.5 and 1.6).

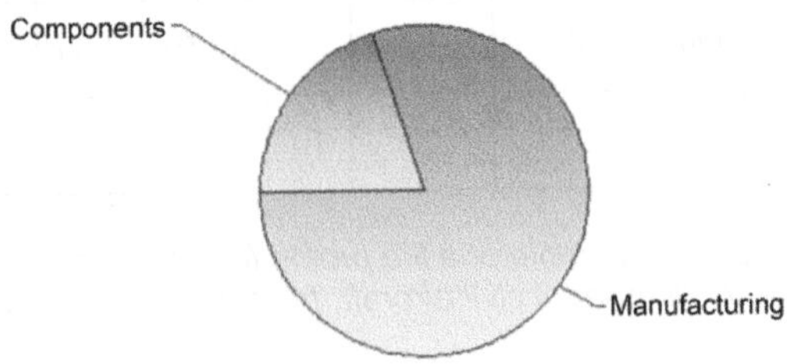

Figure 1.5. *Breakdown for the origins of failures*

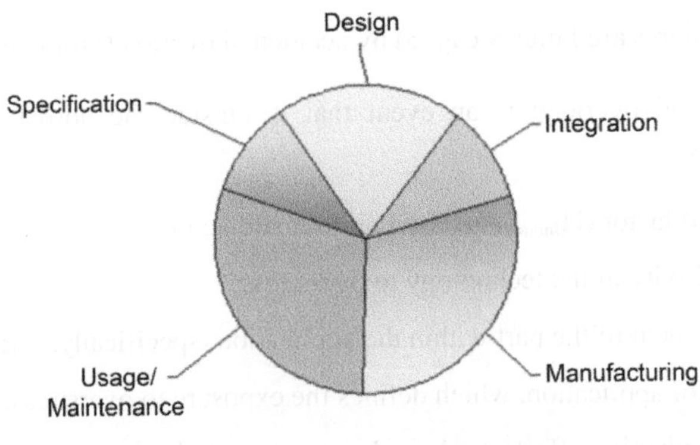

Figure 1.6. *Breakdown of the lifecycle*

This problem has led the FIDES Group to integrate the aspects pertaining to process in the approach of predictive reliability. For this purpose, an approach that allows for the modeling and quantification of the process has been developed. However, this approach is not limited to evaluation: it is mainly a tool to help build reliability. This tool enables us to act throughout all the lifecycle to better control and improve the reliability of products.

The FIDES process factor is based on an audit that includes approximately 150 questions and associated recommendations. Each question is associated with satisfaction criteria that allow us to evaluate the level at which the recommendation is – or is not – applied in a given company.

1.3.8. Sensitivity of FIDES models

Numerous experiments have been carried out by users during and after the development of the method. Therefore, sensitivity analyses have been performed on the main parameters of the models, sometimes by comparing them to other prediction methods (MIL-HDBK-217, RDF2000, etc.). This generally concerns the base failure rates (λ_0) and acceleration factors. As a result, the analysis of acceleration factors has shown the importance of the life profile. To appreciate its influence, the reliability of a digital circuit board of average complexity (about 300 components) was calculated for different typical life profiles taken from the examples provided in the guide (see Figure 1.7). The failure rates obtained increase with the severity of the environment in the following order: desktop computer in France, military radio, armored vehicle, helicopter, military aircraft.

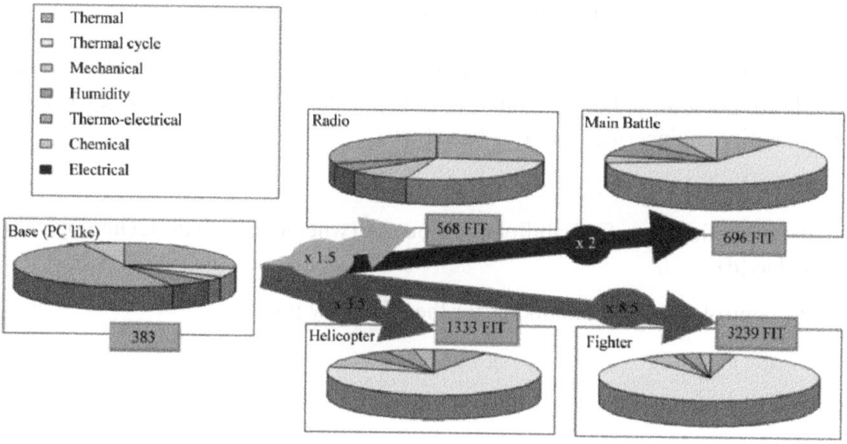

Figure 1.7. *Influence of life profiles (FIT: number of failures per 10^9 hours)*

These results clearly show consideration for the severity of the life profile in the models and their configuration according to the stress level for the same application. Contrary to the MIL-HDBK-217 method which uses a generic factor πE according to the type of environment, the significance of the FIDES model is to provide the share of each environmental contributor at the same time due to the description of all the stressors (mechanical, thermal, thermal cycling, humidity) and their respective level relative to each life profile. Furthermore, it can be seen that the distribution of the contributors changes according to the life profile. In this case, the more severe the application, the more significant the thermal cycling contributor becomes. For further details on the influence of the life profile on FIDES reliability evaluations, refer to [BAY 10].

1.3.9. *Industrial applications*

In many respects, the FIDES methodology remains very similar to existing ones:

– the expected results are failure rates;

– the redundancy or architecture considerations are not taken into account;

– the predicted results can be used in all safety studies.

Nevertheless, the introduction of the FIDES methodology in the company requires changes, particularly because the application of old standards has most often condensed predictive reliability studies into an uncritical application of calculation tables that are rarely well understood. The idea that reliability is not only experienced but also constructed has to be introduced.

The FIDES methodology is developed with this goal in mind. This is why, beyond providing a predictive reliability method, the FIDES methodology is above all a reliability engineering tool. It is based on:

– an explicit approach;

– the identification of reliability contributors (whether physical, technological or process-based);

– the control and audit handbook for the reliability process.

1.4. Validity of reliability studies with FIDES

The evaluation for the predictive reliability of military electronic equipment has historically been carried out using the MIL-HDBK-217 standard, established by the American DoD (Department of Defense). This standard uses a statistical approach to analyze failure data from experiential feedback of electronic components since the 1980s. The MIL-HDBK-217 guide is no longer suitable for today's expectations because it is technically outdated (inadequacy with today's electronic component technologies, very pessimistic reliability predictions, relevance of the results, etc.). Table 1.3 shows that for a sample of different military equipment with various life profiles covering all three armed forces, MIL-HDBK-217F does not give sufficiently relevant evaluations with regard to user feedback, and above all, is very scattered in terms of the amplitude of its deviation. This experiential feedback across 10 different military electronic systems represents the monitoring of 5,200 different electronic boards and 29 million electronic components. Table 1.3 shows the relationship between the MTBF observed in operation (REX: Return from Experience) and the predicted MTBF with MIL-HDBK-217 and FIDES.

Equipment manufacturer	Equipment	MTBF REX/MIL HDBK	MTBF REX/FIDES
A	1	51.8	3.3
	2	10.1	1.1
	3	6.7	0.9
	4	2.8	1.0
B	5	4.8	1.2
C	6	12.4	2.2
D	7	3.1	0.7
	8	3.2	0.8
E	9	2.5	1.3
	10	1.4	0.6

Table 1.3. *Comparison of predicted MTBF MIL HDBK and FIDES versus MTBF REX*

Analysis of Table 1.3 shows that the majority of the forecast reliability results obtained with FIDES are consistent and close to the operational reliability results (REX). For FIDES, some ratios are less than 1, which means that the forecast was better than the reality. This can be explained by the multiple removals (between 1.5 and 2 components removed per failure) during the repairs (these removals were not taken into account in prediction calculations), and due to the limited observation and measurement period as well, which in some cases did not enable us to observe enough failures.

In addition, the feedback from the members of the IMdR FIDES expert group has also shown a good relevance of the forecasts with respect to user feedback on-the-ground. Whatever the field and the type of system studied, and considering the uncertainties inherent to these theoretical forecasts, it is strongly recommended to also integrate, when available, user experience data.

The FIDES methodology, through an approach based on reliability engineering, offers a more realistic prediction than MIL-HDBK-217. This results in the integration of additional parameters related to, among others, the process, the component quality policy, the strength or the application of the RoHS regulation (European Directive 2002/95/EC Restriction of Hazardous Substances). A careful reading of the FIDES 2009 guide, which can be downloaded from www.fides-reliability.org, will provide a number of explanations on the methodology and its application, but also on the implementation of reliability engineering. In addition, there are open access computer tools that have already integrated all the models in the handbook, which can assist with these further studies. The website offers a wide range of support, including a discussion forum. We should not hesitate to ask questions directly and thus obtain a well-founded answer from the moderators who, for the most part, are the experts behind the FIDES method.

The maintenance structure of the method proposes a free experimentation tool (downloadable from the FIDES website) integrating all the models of the FIDES guide. This tool, called the *FIDES Expertool*, enables us to evaluate the predictive failure rates, and has undergone all the validation phases of a professional software tool, which allows the user to have confidence in the results; thus, the user can concentrate on the configuration of the input hypotheses specific to the FIDES methodology. *Expertool* also allows users to implement FIDES without having to enter a list of components. Indeed, the tool has a demo file that is also available via the FIDES website. This demo file is pre-filled with a list of components representative to the scope of the models. This file enables us to become familiar with the different FIDES parameters (life profile, Π_Process, Π_Part-manufacturing, Π_Application, etc.) and to understand the possibilities offered in terms of experimentation (sensitivity studies): influence of the life profile, the variations of the different parameters, etc. As regards to implementation, a user guide is also provided with the tool.

Recommendations to simplify the migration from MIL-HDBK-217 to FIDES are presented in [RIG 14].

1.5. Conclusion

At the initiative of a group of industrialists and the DGA, an alternative to the MIL-HDBK-217 was born. This initiative is not limited to the simple need to predict "relative" failure rates, it also seeks to meet all the current needs of companies in predicting and constructing reliability for the purpose of reliable electronic design.

The FIDES methodology is now operational and is the preferred method for electronic reliability in French weapon programs.

The FIDES method introduces three innovative factors when compared to older reliability standards or databases:

– the need to establish a life profile that is a realistic representation of the situations that the system will actually encounter during its use;

– the contribution of induced factors (also called accidental overloads or overstress): Π_Induced; this factor represents the risk of suffering accidental Mechanical OverStress (MOS), Electrical OverStress (EOS) and Thermal OverStress (TOS). It is evaluated from the sensitivity of the product, its location, its environment of use and the policy of taking into account overstress in the development phase;

– the level of control and construction of reliability in the lifecycle: Π_Process.

The FIDES process tool can be used to varying degrees of depth. The complete audit process, which allows for the identification of critical points, and to establish action plans, obviously requires an initial investment.

FIDES takes better account of all environmental constraints through the life profile when compared with traditional empirical approaches. Therefore, the limitation studies conducted by different users reinforce the need to consider FIDES as a true reliability engineering and construction method. Ideally, it will be integrated into the multiple industrial development processes of an electronic system and will not be satisfied by being a new calculation method in the strictest sense, limited to the use of reliability calculations.

FIDES is considered by its users as a real breakthrough in the field of electronic reliability forecasting. It must be integrated into a TRM (total reliability management) approach, in the same way as the quality processes of the early 1990s were integrated into the TQM (total quality management) approach.

FIDES meets the main needs of electronics manufacturers that are currently seeking an alternative to the MIL-HDBK-217 and RDF2000 handbook (France Telecom - UTE C 80810 or IEC 62380).

Users' feedback show that this method is not limited to the defense field. Today, many manufacturers in the automotive, aerospace and medical fields use FIDES for determining the reliability of their products.

The method is kept up to date through a group within IMdR, wherein all users are invited to participate. In addition, studies have been launched to integrate models related to new technologies (DSM "Deep Sub Micron" components, new power components, etc.) and to update certain parameters that are changing rapidly. These studies will lead to the release of a new version of the FIDES guide in 2021. Finally, a tool is being developed to enable "automatic" sensitivity analyses. Ultimately, this will make it possible to vary certain parameters according to a given interval in order to evaluate the impact of these parameters on the system's reliability result. It will allow, among other things, to evaluate the margins with respect to prohibitive failures. This approach is particularly aimed at safety studies.

1.6. References

[BAU 94] BAUERNSCHUB R., LALL P., "A PoF approach to addressing defect-related reliability", *IEEE/CPMT International Electronics Manufacturing Technology Symposium*, 1994.

[BAY 10] BAYLE F., DAVENEL F., "FIDES : principes théoriques et justification des paramètres", *λμ17 La Rochelle*, 2010.

[BEC 04] BECHTOLD L.E., COMPANY BOEING, "COTS electronics reliability", *IEEE MRQW Conference*, December 2004.

[DEN 97] DENSON W.K., KEENE S.K. "New system reliability assessment methods", PRISM® software tool, RAC Project A06839, 1997.

[FID 09] FIDES, GUIDE FIDES 2009 A, Méthodologie de fiabilité pour les systèmes électroniques, Reference document, UTE C80-811, 2009.

[GUI 04] GUILLET N., GIRAUDEAU M., "FIDES, la nouvelle méthodologie pour la prévision de fiabilité des systèmes électroniques", *λμ14 Bourges*, 2004.

[MAR 05] MARIN J.J., "Experience report on the FIDES reliability prediction method", RAYTHEON, *RAMS Conference*, January 2005.

[MIL 95] MIL-HDBK-217F (+ notice 1 and 2), Reliability prediction of electronic equipment, Military handbook, 1995.

[RIG 14] RIGO J., TOURTELIER D., DAVENEL F., "Ensemble de recommandations pour migrer de la MIL HDBK 2", *Décider dans un monde incertain : enjeu majeur de la maîtrise des risques*, October 21, 2014.

[TOU 10] TOURTELIER D., DAVENEL F., RIGO J., "Influence du profil de vie sur l'évaluation de la fiabilité des systèmes électroniques", *λμ17 La Rochelle*, 2010.

[UTE 00] UTE C 80810 (RDF2000), Recueil de données de fiabilité, France TELECOM, 2000.

[VER 05a] VERGNAULT E., "FIDES : méthodologie de fiabilité pour les systèmes électroniques", *Journal "Actualité composants du CNES"*, vol. 17, p. 6, 2005.

[VER 05b] VERGNAULT E., "FIDES : le retour d'expérience CNES", *Journal "Actualité composants du CNES"*, vol. 18, p. 2, 2005.

2

Reliability in Maritime Transport: Choosing a Container Handling System

The application of a structured RAMS (reliability–availability–maintainability–safety) approach, used in a simple but effective way, is able to provide an overall consistency and demonstrate the achievement of qualitative and quantitative objectives.

Therefore, in the fictitious example proposed, the choice of a container handling system, through a deconstruction of the system into a set of components and their functions, allows us to simply evaluate its reliability (calculation of the mean time between failure [MTBF]), its maintainability (mean time to repair [MTTR]) and its availability, as well as identify the architecture of the handling system (crane with an open kinematic chain with three brakes and type B carabiners), which is able to attain the intended safety objective (risk of a container dropping) of 5×10^{-5} per year.

2.1. Introduction

The production of evidence (justification file) is generally mandatory from a regulatory point of view. This is not always explicit, but nevertheless any industrialist must be aware that they cannot avoid their duty to become an expert on "State of the Art", and still be able to demonstrate it.

The objective of this modest applied case study, totally fictitious and yet also totally plausible, is to simultaneously show both:

Chapter written by Julien RULLIER, Benjamin ECHARD and Ghislaine DELAPAYRE.

For a color version of all the figures in this chapter, see: http://www.iste.co.uk/elhami/applied1.zip.

– the importance of adopting a structured approach with the aim of overall coherence, which is credible from the proof provided (for this, kindly also refer to the following concepts: system engineering – traceability of requirements – quality management – benchmarks such as CMMi or SPice);

– how to handle simply but effectively the ultra-classic methods without losing sight of the objective: to demonstrate qualitative and quantitative objectives concerning the RAMS[1].

2.2. Proposed case study

A company specializing in the design and manufacture of sensitive cargo container handling equipment decided to offer a standard system that could "hoist loads of 38 tons ±2t".

This new system must be able to hoist potentially sensitive cargos, as well as mitigate the risk of the container dropping, either due to their cost or because of the impact they will have on the environment. The mission profile is specified in Table 2.1.

Data	Values
Time of mission per day	3 shifts of 8 h, 1,440 minutes/day
Number of handling cycles per day	Lognormal law of expectation 100 and a variance of 9
Average cycle time assessment[2]	14.4 min
Number of mission days per year	250 days, 6,000 h
Time in charge[3]	1,500 h
Expected lifespan	10 years

Table 2.1. *System mission profile*

1 Reliability, Availability, Maintainability, Safety. Acronym usually used for the discipline of operating safety, thus summarizing the issues covered.

2 Here the term "average cycle time" is a corruption of language. In reality, the number of daily cycles follows a lognormal law, and as such we should really use the term "expectation".

3 During a cycle, the load is not applied to the system until the hoisting is complete. Here, the time under load corresponds to 25% of the total time.

A preliminary hazard analysis (PHA) type analysis was used to identify the potential risks related to handling:

Risk 1: Risk of a container dropping, safety risk (of goods or people)

This risk will have different consequences depending on the nature of the cargo and the environment:

– Example 1: if the container is carrying radioactive waste, its dropping would have serious consequences for the environment (dispersal of radioactivity) and for the safety of the general population. This is therefore a question of personal safety.

– Example 2: if the container is transporting luxury vehicles, its dropping would generate a financial loss for the transporter, thus alluding to the safety of goods.

Risk 2: Risk of untimely stop, risk of unavailability

This risk will affect the efficiency of the handling and transport process. It prevents the system from fulfilling its function, but does not jeopardize the safety of goods or people.

Safety objectives (requirements):

The project manager has included the following requirements in the system specifications:

– Exi-safety: P_{drop} (1 year) < 5×10^{-5}. The residual risk of a container dropping should be limited to an annual probability of 5×10^{-5}.

– Exi-availability: P (be available) > 99%. The intrinsic availability of the system shall be greater than 99%, that is, a probability of 0.99.

– Exi-reliability: the reliability of the system will be calculated by the evaluation of the MTBF[4], which must be at least 4,000 h. This evaluation will be done by neglecting the mechanical components whose failure directly leads to a risk of a container dropping.

– Exi-maintenance[5]: all repairs (except for mechanical elements whose failure leads directly to a risk of a cargo container dropping) must be carried out within a maximum period of 24 h (excluding logistical and organizational time) and the average repair time must be less than 12 h.

4 Mean time between failures (= mean time to failure for components that are irreparable).

5 Note that there are two requirements for maintainability, each of which must be demonstrated.

2.3. Inputs of the RAMS approach

2.3.1. *Presentation of the system*

The risk analysis approach will have to be based on a definition of the system, and different formalisms exist for this: functional block diagrams and structured analysis and design technics (SADT). This definition is not strictly speaking a safety operating procedure (SOP) deliverable, but it is essential to carry out this analysis as it allows us to check with the design team that we have the same understanding of the system and its operation.

The initial design of the system with a service brake is shown in Figure 2.1.

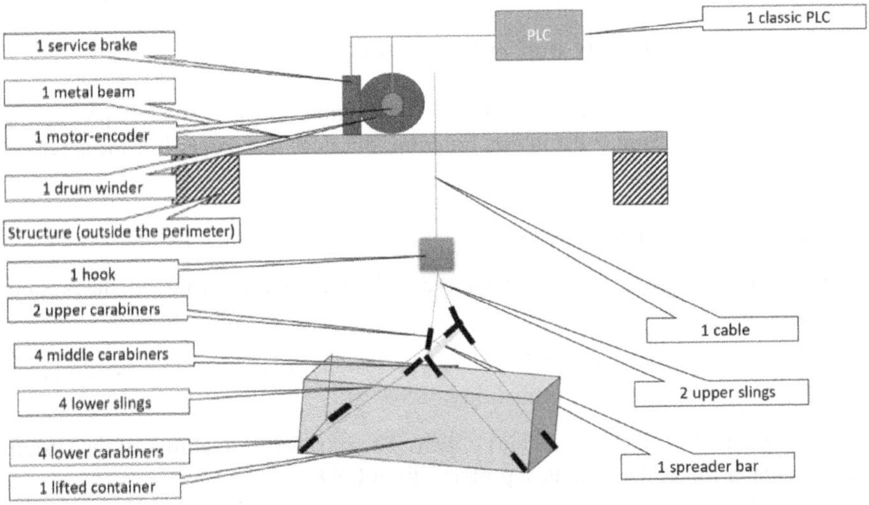

Figure 2.1. *Representation of the handling system with one service brake*

This exercise results in a nomenclature, or bill of materials (BOM), which is specific (see Table 2.2) to a system with one service brake:

– component name;

– a unique identifier for the component (in the example these number from C_1 to C_n);

– the quantification of the number of identical components in the system;

– a description of the technical function performed by the component.

Description	ID	No.	Technical Functions	Proba of Constant Failure (without unit)	Failure Rate (fail/hour)	Proba of Failure over 1 hour of Operation (without unit)	MTBF (hours)	Source or Justification for Quantification	Comments	MTTR (hours)	Intrinsic Unavailability of the Component Group
Service Brake	C1	1	To brake the rotation of the drum to stop it after a nominal movement.	-	1.20E-05	1.20E-05	83,542	NPRD2011 "Brake Assembly, Electric", FG	FG = Fixed Ground	16	0.01915%
Metal Beam	C2	1	To support all the masses of the system and the payload.	1.00E-10	-	-		EUROCODE sizing: compliance with certain "standardized" sizing criteria allows the risk of rupture to be considered negligible and to be modeled at 10-10.	Note: In this case all EUROCODE criteria must be respected.		
Motor Shaft	C3	1	Mechanical link between the motor and the cable drum.	1.00E-10	-	-		EUROCODE sizing: the respect of some "standardized" sizing criteria allows us to consider the risk of negligible rupture and to model it at 10-10.	Note: In this case all the EUROCODE criteria must be respected.		
Motor	C4	1	To ensure the rotation of the winding drum.	-	4.67E-05	4.67E-05	21,404	NPRD2011 "Motor Assembly, Electric", N : 4,67E-5		12	0.05603%
Encoder	C16	1	To acquire information about the rotation of the drum.	-	1.40E-05	1.40E-05	71,480	NPRD2011 "Rotary Encoder" : 13,99E-6	The encoder only makes an acquisition allowing a computer to deduce a speed and/or an acceleration of the drum (and the PLC will trigger the release of the service brake if the speed is judged abnormally high).	6	0.00839%
Classic PLC	C6	1	To manage the rotation movement of the winding drum and thus to manage the raising or lowering of the load + manage the braking of the load (and thus its maintenance); to wind the cables to raise the load.	-	4.67E-05	4.67E-05	21,404	NPRD2011 "Motor Controller"		8	0.03736%

Description	ID	No.	Technical Functions	Proba of Constant Failure (without unit)	Failure Rate (fail/hour)	Proba of Failure over 1 hour of Operation (without unit)	MTBF (hours)	Source or Justification for Quantification	Comments	MTTR (hours)	Intrinsic Unavailability of the Component Group
Winding Drum	C7	1	To wind the cables to raise the load; unwind the cables to lower the load.	1.00E-10	-	-		EUROCODE sizing: the respect of some "standardized" sizing criteria allows us to consider the risk of negligible rupture and to model it at 10-10.			
Cable	C8	1	To ensure the handling of the load; to wind around the reel.	1.00E-10	-	-		EUROCODE sizing: the respect of some "standardized" sizing criteria allows us to consider the risk of negligible rupture and to model it at 10-10.	The safety coefficient of 10 theoretically guarantees a negligible probability of failure (<10-12) but this has no practical application, so we generally limit the reliability by a minimum of 10-10: 1 chance in 10 billion.		
Hook	C9	1	To ensure the handling of the load; to hold the slings.	1.00E-10	-	-		EUROCODE sizing: the respect of some "standardized" sizing criteria allows us to consider the risk of negligible rupture and to model it at 10-10.	Note: In this case all EUROCODE criteria must be respected.		
Spreader Bar	C10	1	To facilitate the mechanical interface between the cables and the slings; to distribute the load.	1.00E-10	-	-		EUROCODE sizing: the respect of some "standardized" sizing criteria allows us to consider the risk of negligible rupture and to model it at 10-10.			
Upper Type B Carabiners	C11	2	To ensure the gripping of the load with a fast and secure attachment.	8.00E-06	-	-		See justifications, reliability results for a 75% confidence level.	Type B: see reliability calculation, Table 2.5 of section 1.3.3.		
Upper Slings	C12	2	To ensure the handling of the load, interface between hook and spreader bar (via carabiners).	1.00E-10	-	-		EUROCODE sizing: compliance with certain "standardized" sizing criteria allows the risk of rupture to be considered negligible and to be modeled at 10-10.	The safety coefficient of 10 theoretically guarantees a negligible probability of failure (<10-12) but this has no practical application, so we generally limit the reliability by a minimum of 10-10: 1 chance in 10 billion.		
Lower Slings	C13	4	To ensure the handling of the load, interface between spreader bar and container (via carabiners).	1.00E-10	-	-		EUROCODE sizing: the respect of some "standardized" sizing criteria allows us to consider the risk of negligible rupture and to model it at 10-10.	The safety coefficient of 10 theoretically guarantees a negligible probability of failure (<10-12) but this has no practical application, so we generally limit the reliability by a minimum of 10-10: 1 chance in 10 billion.		
Lower Type B Carabiners	C14	4	To ensure the gripping of the load with a fast and secure attachment.	1.00E-10	-	-		See justifications, reliability results for a 75% confidence level.	Type B: see reliability calculation, Table 2.5 of section 1.3.3.		
Middle Type B Carabiners	C15	4	To ensure the gripping of the load with a fast and safe attachment.	1.00E-10	-	-		See justifications, reliability results for a 75% confidence level.	Type B: see reliability calculation, Table 2.5 of section 1.3.3.		

Table 2.2. *Component reliability data*

2.3.2. *Component reliability data*

Table 2.2 shows that we have already started the reliability analysis by stating data concerning reliability and maintainability:

– probability values or failure rates;

– MTBF, if applicable;

– quantification source (example here of NPRD2011 [NPR 11]);

– values for mean time to repair (MTTR).

However, not all components will be found in the reliability data[6], and it will sometimes be necessary for some of them, when they are specific mechanical components:

– to guarantee their reliability through the EUROCODE: the respect of certain "standardized" sizing criteria allow us to consider the risk of negligible breakage and to model it at 10^{-10} for the slings, the beam, the motor axis, the spreader bar, etc.;

– to have a finer approach, in order to determine the predicted unreliability by taking into account the stresses really undergone by the component and its capacity to resist it. The demonstration proposed below relates to carabiners.

2.3.3. *Reliability of carabiners over time*

In order to introduce the approach of reliability over time, also called durability, we consider in this fictitious example two models of carabiners, bought in the market, and for which we want to ensure the behavior over time in view of ageing since we do not know their sizing criteria or safety margins. We assume that these carabiners are correctly sized from a static point of view.

The reliability approach used here is divided into two phases: an analysis phase that aims to identify and model the failure of physical mechanisms, and then a reliability estimation phase through testing [DEL 17].

6 It is interesting to note here that even if there are many reliability databases, it is always difficult (even illusory) to guarantee that the data used for the estimation are representative of the intended use. Nevertheless, this quantitative approach is recognized and quite relevant, especially for making comparisons (between multiple design choices).

2.3.3.1. *Phase 1: analysis of the failure of physical mechanisms and damaging factors*

In this application, the carabiners will be subjected to repeated loading cycles. This type of loading progressively damages the material and can lead to the development of a crack. This failure mechanism is called fatigue.

Fatigue is generated by two damaging factors:

– the number of handling cycles;

– the stress related to the weight of the load.

2.3.3.1.1. Mission profiles of damaging factors

To verify the fatigue life of the carabiner, it is necessary to know the mission profiles of the two damaging factors. The mission profile describes the evolution of the factor during the lifetime of the component. It can be a deterministic (fixed) value or a random variable.

Table 2.1 gives the following information for the number of handling cycles:

– number of cycles per day: lognormal distribution with an expectation of 100 and a variance of 9;

– operating time per year: 250 days per year for 10 years.

In the following, we note $N10$, the random variable corresponding to the number of cycles over 10 years. It follows a lognormal with an expectation of $100 \times 250 \times 10$ and a variance of $9 \times (250 \times 10)^2$, that is, a mean of $m_{lnN10} \approx 12.429$ and a standard deviation of $s_{lnN10} \approx 0.03$.

For the second damaging factor, that is, the stress related to the weight of the load, it is necessary to calculate the force on each carabiner. For this fictitious example, inertias are considered negligible (i.e. a very slow and smooth movement speed).

The force F_S for an upper carabiner is:

$$F_S = \frac{M\,g}{2\cos(u/2)} \approx 227 \text{ kN} \quad [2.1]$$

where:

– M is the mass of the container, which is 40 tons. It will be assumed in the following that the container is perfectly balanced and the masses of the slings and the spreader bar are negligible in relation to the container;

– g is the acceleration of gravity (9.81 m·s^{-2});

– u is the angle between the two upper slings, here 60°.

The force F_I taken by each lower and middle carabiner is:

$$F_I = \frac{M\,g}{4\cos(v/2)} \approx 139 \text{ kN} \qquad [2.2]$$

where v is the angle between the two lower slings, here 90°.

It can be seen that the upper carabiners are the most stressed.

In the following, it is assumed that the stress level S in the carabiner is proportional to the force F:$S = \alpha\,F$.

2.3.3.1.2. Model for fatigue damage

The assumed fatigue model for the carabiner is the Basquin model:

$$N\,S^b = \text{constant} \qquad [2.3]$$

where N is the number of cycles up until failure and b is the fatigue slope. The latter, according to the supplier, is equal to 8. Examples of fatigue models are given in [BER 17] and [MIL 03].

Due to the proportionality between stress and force, the model is rewritten as:

$$N\,F^b = \text{constant} \qquad [2.4]$$

2.3.3.2. *Phase 2: estimating reliability through testing*

The stress–strength method [DEL 17] (Figure 2.2) is used to estimate the probability of the carabiners failing. This method is based on the comparison of two statistical distributions:

– the “stress“ distribution, which represents the damage generated by usage over the intended lifetime and takes into account the uncertainties related to customer use;

– the "strength" distribution, which represents the tolerable damage of the carabiner and takes into account the uncertainties related to its manufacture and materials.

It is important to note that these two statistical distributions must be defined on a common x-axis. In this case, the axis is the number of cycles under F_k ($k = S$) or I for the upper, middle and lower carabiner.

Reliability is intended to quantify the probability of failure, that is, the probability that the damage tolerated by the carabiner is less than the damage generated by the customer.

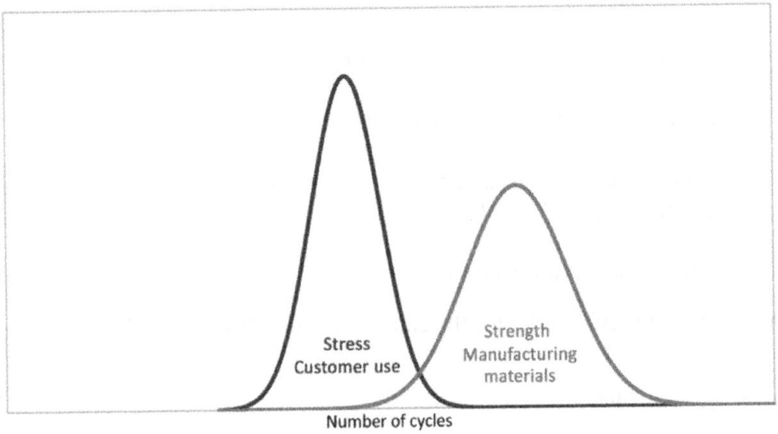

Figure 2.2. *Stress–strength method*

2.3.3.2.1. Modeling stress

Given the definition of the x-axis above, the stress distribution corresponds directly to the distribution of the random variable $N10$. As a reminder, it follows a lognormal with an expectation of $100 \times 250 \times 10$ and a variance of $9 \times (250 \times 10)^2$, that is, the mean $m_{lnN10} \approx 12.429$ and the standard deviation $s_{lnN10} \approx 0.03$.

2.3.3.2.2. Modeling strength

As the strength of the carabiners is not known, it is necessary to perform failure tests to estimate it. The test protocol is as follows:

– sinusoidal stress cycle under repeated tension (Figure 2.3) of $F_{test} = 250$ kN;

– stopping criterion: appearance of a crack;

– 10 carabiners tested per model randomly taken from the supplier's production.

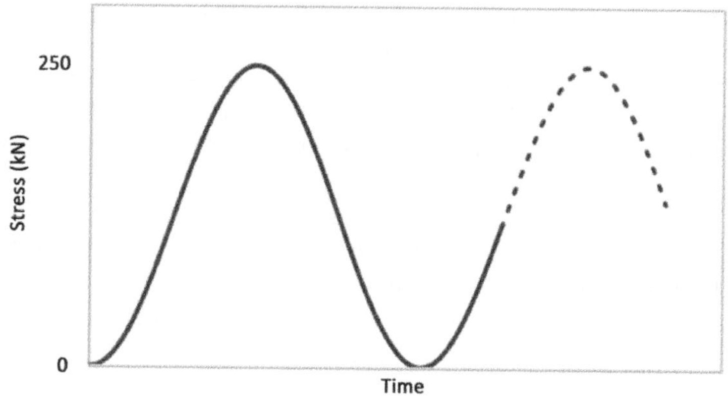

Figure 2.3. *Stress cycle during testing*

The test is performed at a higher stress level than when in use in order to precipitate failure. This is called an accelerated test. From the fatigue damage model, it is possible to calculate the number of cycles N_{0k} under F_k ($k = S$ or I for upper, middle and lower) producing damage equivalent to that experienced in the test:

$$N_{0k} = N_{test}\left(\frac{F_{test}}{F_k}\right)^b \quad [2.5]$$

Table 2.3 gives these numbers of cycles until failure for each carabiner tested.

#	Carabiner A			Carabiner B		
	N_{test}	N_{0S}	N_{0I}	N_{test}	N_{0S}	N_{0I}
1	138,346	304,189	1.54×10^7	151,981	334,166	1.69×10^7
2	160,125	352,073	1.78×10^7	175,071	384,935	1.95×10^7
3	174,359	383,369	1.94×10^7	174,345	383,339	1.94×10^7
4	134,566	295,876	1.50×10^7	148,686	326,921	1.65×10^7
5	166,354	365,769	1.85×10^7	163,500	359,493	1.82×10^7
6	142,188	312,634	1.58×10^7	169,510	372,707	1.88×10^7
7	147,913	325,222	1.64×10^7	157,358	345,989	1.75×10^7
8	154,891	340,564	1.72×10^7	168,055	369,509	1.87×10^7
9	164,109	360,832	1.82×10^7	177,429	390,119	1.97×10^7
10	179,563	394,812	2×10^7	185,754	408,424	2.07×10^7

Table 2.3. *Failure test results*

The dispersion of the cohesion cycle count for each model can be modeled by a statistical distribution. The Johnson rank method is used to apply a lognormal distribution to the N_{0k} values. The distribution selected to estimate reliability corresponds to a 75% confidence level. This means that the selected distribution has a 75% chance of being less than the real distribution. These parameters are outlined in Table 2.4.

		Carabiner A	Carabiner B
N_{0I}	Expectation	1.68×10^7	1.82×10^7
	Variance	$(0.11 \times 1.68 \times 10^7)^2$	$(0.078 \times 1.82 \times 10^7)^2$
N_{0S}	Expectation	333,056	359,270
	Variance	$(0.11 \times 333{,}056)^2$	$(0.078 \times 359{,}270)^2$

Table 2.4. *Strength distribution parameters for a 75% confidence level*

2.3.3.2.3. Ten-year reliability estimate

In the case of two lognormal distributions, it can be shown that the probability of failure P_f at 10 years is equal to:

$$P_f = \Phi\left[\frac{-\left(m_{lnN0k} - m_{lnN10}\right)}{\sqrt{s_{lnN0k}^2 + s_{lnN10}^2}}\right] \quad [2.6]$$

where Φ is the distribution function of the reduced centered normal distribution.

Table 2.5 gives the P_f values for each carabiner model. Carabiner B has the lowest probability of failure.

P_f to 10 years	Carabiner A	Carabiner B
Middle and lower carabiner	$\ll 10^{-10}$	$\ll 10^{-10}$
Upper carabiner	6.7×10^{-3}	8×10^{-6}

Table 2.5. *Summary of reliability results for a 75% confidence level*

The carabiner failure probabilities are added to Table 2.2 and the system's RAMS objectives can now be evaluated.

2.4. Assessment of the system's RAMS

The following paragraphs will evaluate the safety operation procedures by working successively through all the aspects of RAMS:

– *reliability* through assessing the overall MTBF of the system and comparing it to the target;

– *availability* by calculating the availability of the complete system;

– *maintainability* by evaluating the global MTTR of the system and comparing it to the objective;

– *safety* by analyzing in detail the modes of failure that can generate a safety risk (FMECA), and by evaluating the probability of occurrence of this risk according to all the different possible scenarios (fault tree).

2.4.1. *Reliability assessment*

As stated in the objective, the MTBF is evaluated here without taking into account the mechanical components (carabiners, slings, etc.) since they have a negligible impact on the result, due to their very low probabilities of failure. The following assumptions are made:

– the system is in series, that is, the failure of one component leads to the failure of the system;

– the components considered in the MTBF calculation are assumed to follow exponential laws as per their failure rates λ provided in Table 2.2.

For these assumptions, the MTBF of the system is the inverse of the sum of the failure rates λi of its components, which is:

$$MTBF_{system} = \frac{1}{\sum_{i=1}^{n} \lambda i} \qquad [2.7]$$

For the case study, the MTBF of the system is:

$$1/(1.2 \times 10^{-5} + 4.67 \times 10^{-5} + 1.4 \times 10^{-5} + 4.67 \times 10^{-5}) = 8{,}375 \text{ h}$$

The MTBF target of 4,000 h has more than been met, as the system is likely to fail, on average, every 8,375 h.

2.4.2. *Assessment of the intrinsic availability*

The formula for evaluating availability[7] is:

$$Availability_{average} = \frac{MTBF}{MTBF+MTTR} \quad [2.8]$$

We then calculate for each component its availability and we consider that a system is available if all its components are available simultaneously, which means that:

$$Availability_{System} = \prod_{i=1}^{n} Availability_i \quad [2.9]$$

In our example, we obtain an availability of 99.88%: the objective of 99% is thus reached!

COMMENT 1.– THE OPTIMIZATION OF INTRINSIC AVAILABILITY

If the objective had not been met, it would have been possible to optimize corrective maintenance times by improving repairability and its evaluation, and then the failure rates of the components, by limiting their failure frequency.

COMMENT 2.– THE OPTIMIZATION OF LOGISTICAL SUPPORT

This so-called intrinsic availability assessment does not take into account logistical times. All support elements are considered to be perfect (supply time, availability of spare parts, maintenance personnel, return to service time, etc.). The assessment of operational availability, which takes into account these logistical times, is carried out within the framework of "integrated logistical support"[8] activities. The formula for assessing operational availability is given as:

$$Availability_{operational} = \frac{MTBF}{MTBF+MTTR+ Time_{logistics}} \quad [2.10]$$

Let us imagine for a moment that it simply takes 2 supply days for each component; we would then have an operational availability that would fall to 98.74%.

The purpose of the Integrated Logistics Support (ILS) is to design, at the same time as product design, the logistic support system that will guarantee and optimize its operational availability. For example, an effort will be made to store spare parts

7 Assumption of a constant failure rate λ and repair rate μ, or having converged to a constant value. This is referred to as asymptotic availability.

8 See SLI and MIL STD 1388 [MIL 83] or more recently ASD/AIA S3000L [ASD 15].

for components that often break down, and whose supply time would be long, thereby rendering the system unavailable.

It is also possible to introduce here the "cost" parameter into the model and thus perform "asset management", that is, investment control. Indeed, the logistical support model will make it possible to choose when the spare parts should be stored, identifying those that will increase availability the most, at the lowest cost.

2.4.3. *Maintainability assessment*

We can propose an evaluation of the average repair time by the following formula (it is easy to understand that the $MTTR_{system}$ is a weighting of the MTTRs by their failure rates: indeed, if a failure rate has a higher value, the component will "weigh" more heavily in the average):

$$MTTR_{system} = \frac{\sum_{i=1}^{n}(\lambda_i \times MTTR_i)}{\sum_{i=1}^{n}\lambda_i} \qquad [2.11]$$

In our case, we obtain an $MTTR_{system}$ = 10.13 h.

The two maintainability objectives are thus met:

– the average repair time is less than 12 h;

– no repair time is greater than 24 h.

2.4.4. *Safety assessment*

2.4.4.1. *Failure mode and effects analysis (FMEA)*

The method proposed to carry out an assessment on the impact of possible failures is the FMEA method. This method has been theorized and applied since the 1960s and is currently described in the standard NF EN 60812 [NFE 06]. It is very often required for quality standards, for example, those of the automotive industry IATF/ISO 16949 or VDA.

FMEA is an inductive method: the effect of the cause of the failure is analyzed.

It is also generally referred to as FMECA (mode of failure, effects and criticality analysis). In this case, in addition to the criteria of occurrence (causes) and severity (effects), we want to highlight the acceptable or unacceptable nature of the severity/occurrence[9] combination (Table 2.6).

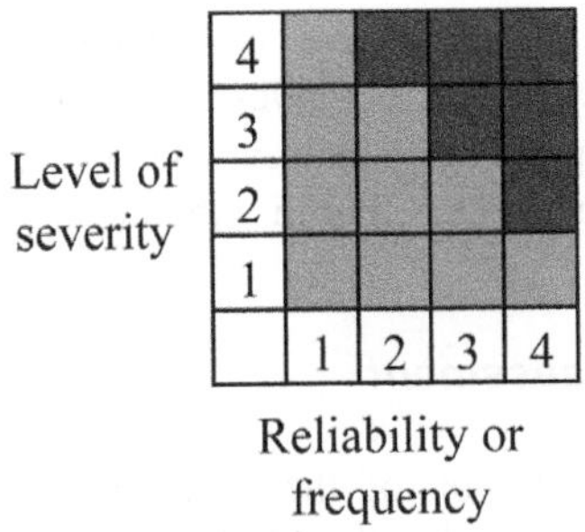

Table 2.6. *Example of a risk criticality/acceptability table*

There are other methods than the FMEA or FMECA that are able to obtain a global vision of the different ways in which a product may fail so as to generate specific types of risk assessment. We can cite as examples:

– design review based on mode of failure (DRBFM) was developed at TOYOTA and is based on the analysis of the impact of changes between an old proven design and a new one; this method is widely used in the automotive industry and recognized in the United States [SOC 13] and Europe [AAI 14];

– reliability block diagram (RBD) makes it possible to represent a system as a network of components whose architecture gives information on the ways it can fail: this method has the advantage of being able to introduce the notion of conditions in the occurrence of a failure (conditional probabilities as part of a fault tree) and to keep a "natural" physical representation of the architecture. The understanding quickly becomes limited in the case of very complex systems.

The results of the method applied to the container handling system are presented in Table 2.8. The meaning of the different columns is given as follows:

– label: recall a component's name;

– Id: identification number of the component;

– Nbr: number of components having this function in the system;

---

9 Acceptability thresholds, corresponding to residual risk tolerances, are defined either through regulations or standards (IEC 61508, EN 50129, ISO26262 or ISO13849), or by the company's General Management, as they lead to the acceptance of a risk, which will have an impact on the brand image, the cost of the guarantee, or even the cost of insurance and the legal risk.

– technical functions: function (use) of the component;

– proba of failure (without unit): value for probability in cases where mechanical components are likely to have constant probabilities of failure;

– failure rate (fail/hours): value of λ in the case of components following an exponential law;

– proba (mission time): evaluation for the probability of failure during a given mission time (here 1 h, sometimes over the component's lifetime);

– mode of failure: way in which the component is led to fail, to not fulfill its function correctly (the function is then, according to the cases "absent", "lost", "untimely", or "degraded");

– cause: physical origin of the failure (mechanical rupture, seizure, etc.);

– distribution by mode of failure: distribution in % of a failure rate among the different possible modes of failure. For an electric motor with three modes of failure and $\lambda = 4.67 \times 10^{-5}$ fail/h:

Mode failure	Mode distribution	Mode failure rate
Loss/absence	10%	4.67×10^{-6} fail/h
Intempestive	10%	4.67×10^{-6} fail/h
Erroneous	80%	3.74×10^{-5} fail/h

Table 2.7. *Distribution by mode of failure*

– proba (mission time): we evaluate the probability of failure for a given mission time, generally 1 h of operation, or, in this case the effective annual probability (1,500 h);

– system effect: the risk is evaluated according to the severity of its effect, by answering the question "what stake is lost or disturbed by the occurrence of this risk?" Here, we have adopted a binary severity scale: either the risk impacts availability (by decreasing it), or the risk impacts safety (by increasing the probability of occurrence of a safety risk such as a cargo container dropping);

– comments: this field can be used to justify the link between cause and effect, for example, to explain how the failure of a mechanical element does not directly lead to a container dropping, since built-in redundancies are in place in the system.

1	Description ENG	Description FR	ID	No.	Technical Functions	Proba (Mission Time)	Mode of Failure	Cause	Failure Mode Distribution	λ of Mode (Fail Time)	Effective Annual Probability	System Effect	Severity (Safety/Avail)	Comments
2	Service Brake	Frein de service	C1	1	To brake the rotation of the drum to stop it following a nominal movement.	1.78E-02	Loss	Wear and tear		0.00E+00		The movement of the load cannot be stopped and it will eventually drop.	Safety	
3	Service Brake	Frein de service	C1	1	To brake the rotation of the drum to stop it following a nominal movement.	1.78E-02	Absent	Wear and tear	15%	1.80E-06	2.67E-03	The movement of the load cannot be stopped and it will eventually drop.	Safety	
4	Service Brake	Frein de service	C1	1	To brake the rotation of the drum to stop it following a nominal movement.	1.78E-02	Degraded	Wear and tear	60%	7.18E-06	1.07E-02	The movement of the load cannot be stopped and it will eventually drop.	Safety	After a certain threshold of wear and tear, the degraded braking corresponds to no braking. Below the threshold it is not a safety issue.
5	Service Brake	Frein de service	C1	1	To brake the rotation of the drum to stop it following a nominal movement.	1.78E-02	Untimely	Break	15%	1.80E-06	2.67E-03	It is impossible to continue the movement, to lift or lower the load.	Availability	
6	Metal Beam	Poutre métallique	C2	1	To support all the masses of the system and the payload.	1.00E-10	Loss	Rupture	100%	No object	1.00E-10	The beam does not support the load anymore and it will drop.	Safety	
7	Motor shaft	Axe moteur	C3	1	Mechanical link between the motor and the cable drum.	1.00E-10	Loss	Rupture	100%	No object	1.00E-10	The axle can no longer support the load and it will drop.	Safety	
8	Motor	Moteur	C4	1	To ensure the rotation of the winding drum.	6.77E-02	Absent	Break grippage	10%	4.67E-06	6.77E-03	It is impossible to continue the movement, to raise or lower the load.	Availability	
9	Motor	Moteur	C4	1	To ensure the rotation of the winding drum.	6.77E-02	Degraded	Wear and tear	10%	4.67E-06	6.77E-03	It is impossible to continue the movement, to raise or lower the load.	Availability	Above a certain threshold, otherwise no effect.

1	Description EN0	Description FR	ID	No.	Technical Functions	Proba (Mission Time)	Mode of Failure	Cause	Failure Mode Distribution	λ of Mode (Fail Time)	Effective Annual Probability	System Effect	Severity (Safety/Avail)	Comments
12	Encoder	Codeur	C16	1	#N/A	2.08E-02	Loss	Loss of electrical connection	10%	1.40E-06	2.08E-03	No speed detected by the system although there is one and it may be too high.	Safety	This defect could be diagnosed by the PLC but not accurately.
13	Encoder	Codeur	C16	1	To acquire information about the rotation of the drum.	2.08E-02	Absent	Loss of electrical connection	10%	1.40E-06	2.08E-03	No speed detected by the system when there is one and it may be too high.	Safety	This defect could be diagnosed by the PLC but not accurately.
14	Encoder	Codeur	C16	1	To acquire drum rotation information.	2.08E-02	Degraded	Signal degradation	80%	1.12E-05	1.66E-02	Incorrect speed evaluation and risk of overspeeding.	Safety	This defect could be diagnosed by the PLC but not accurately.
15	Classic PLC	Automate classique (PLC)	C6	1	To manage the rotation movement of the winding drum and thus manage the raising or lowering of the load + manage the braking of the load (and thus its maintenance); manage the rotation movements of the drum.	6.77E-02	Loss	Internal failure leading to a lack of brake control	10%	4.67E-06	6.77E-03	No braking control by the PLC and risk of load dropping.	Safety	There is also the risk of losing motion but this is less serious (available).
16	Classic PLC	Automate classique (PLC)	C6	1	To manage the rotation movement of the winding drum and thus manage the raising or lowering of the load + manage the braking of the load (and thus its maintenance); manage the rotation movements of the drum.	6.77E-02	Absent freinage	Internal failure leading to a lack of brake control	5%	2.34E-06	3.38E-03	No braking control by the PLC and risk of load dropping.	Safety	There is also the risk of losing motion but this is less serious (available).
17	Classic PLC	Automate classique (PLC)	C6	1	To manage the rotation movement of the winding drum and thus manage the raising or lowering of the load + manage the braking of the load (and thus its maintenance); manage the rotation movements of the drum.	6.77E-02	Erroneous movement	Internal failure leading to untimely motor control	5%	2.34E-06	3.38E-03	Any type of erroneous movement possible (up instead of down, moving instead of braking, etc.) and therefore risk of the load dropping.	Safety	Unless there is evidence to the contrary, it must be assumed that there is a non-negligible probability that a PLC will launch an error maneuver. This can be due to the HW and/or SW.
18	Classic PLC	Automate classique (PLC)	C6	1	To manage the rotation movement of the winding drum and thus manage the raising or lowering of the load + manage the braking of the load (and thus its maintenance); manage the rotation movements of the drum.	6.77E-02	Untimely movement		5%	2.34E-06	3.38E-03	Untimely lifting: risk of being caught and the load dropping. Untimely lowering: shock on impact with the ground.	Safety	Unless there is evidence to the contrary, it must be assumed that there is a non-negligible probability that a PLC will launch an error maneuver. This can be due to the HW and/or SW.
19	Classic PLC	Automate classique (PLC)	C6	1	To manage the rotation movement of the winding drum and thus manage the raising or lowering of the load + manage the braking of the load (and thus its maintenance); manage the rotation movements of the drum.	6.77E-02	Absence of movement	Internal failure of the PLC	75%	3.50E-05	5.08E-02	Impossible movement, no handling.	Availability	The load remains blocked at the bottom.
10	Drum Reel	Tambour enrouleur	#N/A	#N/A	To wind the cables to raise the load; to unwind the cables to lower the load.	#N/A	Loss	Rupture	100%	No object	#N/A	The breakage of the axis or of the connection with the axis of the drum will lead to the untimely unwinding of the drum and thus the load dropping.	Safety	

Description ENG	Description FR	ID	No	Technical Functions	Proba (Mission Time)	Mode of Failure	Cause	Failure Mode Distribution	λ of Mode (Fail Time)	Effective Annual Probability	System Effect	Severity (Safety/Avail)	Comments
Cable	Câble	C8	1	To ensure the handling of the load. Wind around the reel.	1.00E-10	Loss	Rupture	100%	No object	1.00E-10	The breakage of the rope directly leads to the load dropping.	Safety	
Hook	Crochet	#N/A	#N/A	To ensure the handling of the load. Wind around the reel.	#N/A	Loss	Rupture	100%	No object	#N/A	The breakage of the hook directly leads to the load dropping.	Safety	
Spreader Bar	Palonnier	C10	1	To facilitate the mechanical interface between the cables and the slings, to distribute the load.	1.00E-10	Loss	Rupture	100%	No object	1.00E-10	The breakage of the spreader bar directly leads to the load dropping.	Safety	
Upper Type B Carabiners	Mousquetons supérieurs type B	C11	2	To ensure the gripping of the load with a fast and secure attachment.	8.00E-06	Loss	Rupture	100%	No object	1.60E-05	The breakage of one of the two upper carabiners leads to a sudden unbalance of the spreader bar, the container and thus the load.	Safety	
Upper Slings	Elingues supérieures	#N/A	#N/A	To ensure the handling of the load; the interface between hook and spreader bar (via carabiners).	#N/A	Loss	Rupture	100%	No object	#N/A	The breakage of one of the two upper slings leads to a sudden unbalance of the spreader bar, the container and therefore the load.	Safety	
Lower Slings	Elingues inférieures	#N/A	#N/A	To ensure the handling of the load interface between spreader bar and container (via carabiners).	#N/A	Loss	Rupture	100%	No object	#N/A	The breakage of one of the two lower slings leads to a sudden unbalance of the spreader bar, the container and therefore the load, but not of the load dropping.	Availability	It would take the simultaneous or consecutive rapid breakage of at least 2 of the 4 slings to directly cause the load to drop. However, this breakage indicates an abnormal phenomenon (overload due to imbalance, premature wear and tear, fatigue due to repeated loading, etc.) and it will be essential to stop operations, inspect and change (maintain) that part of the equipment: other carabiners, slings, etc.
Lower Type B Carabiners	Mousquetons inférieurs type B	C14	4	To ensure the gripping of the load with a fast and secure attachment.	1.00E-10	Loss	Rupture	100%	No object	4.00E-10	The breakage of one of the four lower snap hooks does not lead to a sudden imbalance.	Availability	It would take the simultaneous or consecutive rapid breakage of at least 2 of the 4 carabiners to directly cause the load to drop. However, this breakage indicates an abnormal phenomenon (overload due to imbalance, premature wear and tear, fatigue due to repeated loading, etc.) and it will be essential to stop operations, inspect and change (maintain) that part of the equipment: other carabiners, slings, etc.
Middle Type B Carabiners	Mousquetons intermédiaires type B	C15	4	To ensure the gripping of the load with a fast and safe attachment.	1.00E-10	Loss	Rupture	100%	No object	4.00E-10	The breakage of one of the four intermediate carabiners does not lead to a sudden imbalance.	Availability	It would take the simultaneous or consecutive rapid breakage of at least 2 of the 4 carabiners to directly cause the load to drop. However, this breakage indicates an abnormal phenomenon (overload due to imbalance, premature wear and tear, fatigue due to repeated loading, etc.) and it will be essential to stop operations, inspect and change (maintain) that part of the equipment: other carabiners, slings, etc.

Table 2.8. *FMEA*

2.4.4.1.1. Conclusions and limitations

We can see that the FMEA allows us to make an interesting exhaustive assessment of the modes of failure, and demonstrate which ones affect safety and which ones are, on the contrary, simply contributors to reliability and unavailability.

If an FMECA is performed, the design can be oriented to meet the acceptability thresholds. In the case of our example, the FMEA will be an essential prerequisite for the next step: modeling safety risks via fault trees. Indeed, we will use the modes of failure identified in the FMEA to model these as the basic events for the fault tree.

The rigor of the overall methodological approach requires that the same wording be used, and even that traceability be ensured between stages. Therefore, the causes of the modes of failure in the FMEA must be the same as the causes of the basic events in the fault trees.

2.4.4.2. *Quantification for the risk of a container dropping*

The target risk for a container dropping is 5×10^{-5} over 1 year.

The method used to quantify this risk is a fault tree analysis. It is a deductive analysis method that consists of identifying the combinations of events that can lead to the feared top event (i.e. the risks associated with a cargo container dropping).

The main steps when building a fault tree are as follows:

1) start from the feared top event that corresponds to the top of the fault tree;

2) imagine the possible middle events that could lead to the top event;

3) consider each middle event as a new vertex event (scenario);

4) imagine the possible causes of each event at the considered level;

5) progressively go down the fault tree to the base events.

In our study, the modeling of fault trees is performed using the software Arbre Analyst V2.3.2 [CLE 14] in which the basic elements or gates shown in Table 2.9 are found.

Modeling with Arbre Analyst	Description
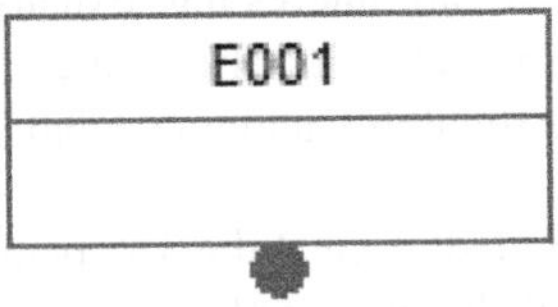	Basic event: the lowest level event for which the probability of occurrence or reliability data is available.
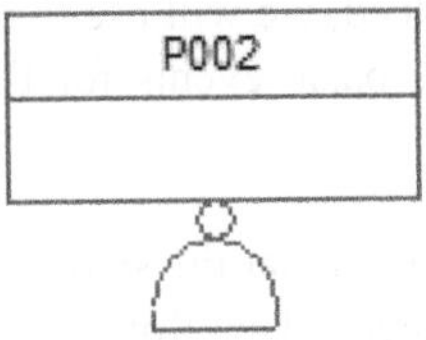	AND gate: the output event "P002" appears if all the input events appear.
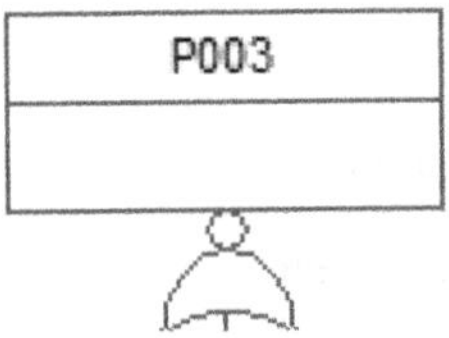	OR gate: the output event "P003" appears if at least one of the input events appear.
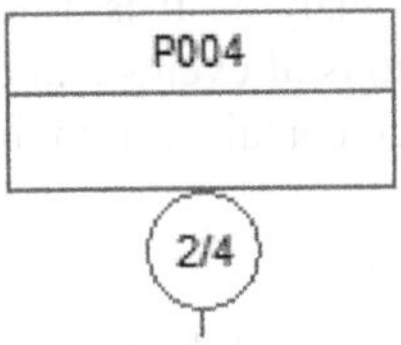	Logic gate of type *k/N*: the output event "P004" appears if *k* input events occur out of *N* possible input events.
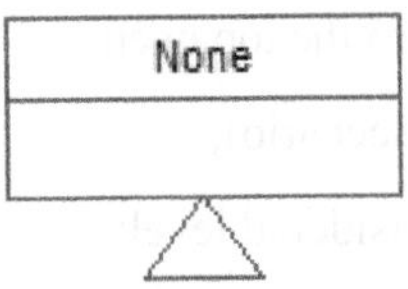	TRANSFER gate: refers to another fault tree.

Table 2.9. *Modeling with Arbre Analyst V2.3.2*

2.4.4.2.1. Modeling assumptions

– *Assumption 1*: The only hazardous event modeled in this study is the risk of a container dropping. The risk scenario for capping (exceeding the high threshold) is not considered.

– *Assumption 2*: The service brake or the safety brake is sufficient to maintain the load (see FMECA).

– *Assumption 3*: The container drops during the failure:

- one upper carabiner out of every two;
- two out of four middle carabiners;
- two out of four lower carabiners (see FMEA section 2.4.4.1).

– *Assumption 4*: Loss of external power supply is not considered in the study. In case of loss of external power supply, the system is designed to ensure the control of the load being handled.

– *Assumption 5*: Hidden failures are excluded from this study because these safeties are checked before each cycle.

– *Assumption 6*: The two safety encoders are for different technologies (see FMEA section 1.4.4.1).

2.4.4.2.2. Mission profile

The annual operating time of the bridge under load is 1,500 hours (see Table 2.1).

2.4.4.2.3. Calculation for the probabilities of failure of basic events

Mechanical components

All the mechanical parts of the handling equipment except the carabiners have been designed using the Eurocode standards with a safety coefficient equal to 2.2 according to finite element modeling (FEM).

For these parts, the probabilities of failure are calculated from the stress/strength method with a dispersion of 8% for the average strength value (a tightly controlled process) and a dispersion of 10% for the average stress value. The probability of failure is hence 10^{-10}.

In order to determine the probability of failure for commercially available carabiners, failure tests according to the bridge use profile were performed (see section 1.3.3.2). Probabilities less than 10^{-10} were limited to 10^{-10} in the modeling because a lower probability seems unrealistic.

The probabilities of carabiner failure are given in Table 2.5.

Other components

The probability of failure of components, other than mechanical components, is automatically calculated by the Arbre-Analyst software according to the following formula:

$$Pf = 1 - \int_0^T e^{-\lambda t}.dt \qquad [2.12]$$

These basic failure rates are given in Table 2.2.

2.4.4.2.4. Quantification for the feared event of a "container dropping"

Bridge with one service brake

The feared event of a "container dropping" is modeled by a fault tree to quantify the likely occurrence of this event (see Figure 2.1).

The risk of a "container dropping" is evaluated for two types of carabiners (A and B), and for each one, the minimum decisive cuts leading to the feared event of a "container dropping" are specified. A minimum cut of order 1 means that equipment failure directly led to the feared event. A minimum cut of order 2 means that a simultaneous failure (i.e. a double failure) by two pieces of equipment is necessary to lead to the feared event.

The scenarios that can lead to the dropping of a container are as follows:

– failure of the steel beam;

– breakage of the kinematic chain;

– loss of electrical torque.

The fault trees are presented in section 1.4.4.2.5.

Assessment for the risk a container dropping with type A carabiners

With a bridge equipped with a service brake, the annual probability of a container dropping with Type A carabiners is 1.9×10^{-2}.

Table 2.10 shows the minimum decisive cuts.

Rank	Annual probability	Contribution (%)	Event 1	Event 2	Cutting order
1	6.7×10^{-3}	34.9	E049: carabiner failure up 2		1
2	6.7×10^{-3}	34.9	E048: carabiner failure up 1		1
3	3.7×10^{-3}	19.2	E031: electrical motor failure	E099: classic PLC failure	2
4	1.1×10^{-3}	5.9	E031: electrical motor failure	E100: encoder failure	2
5	0.9×10^{-3}	5.1	E031: electrical motor failure	E105: service brake failure	2

Table 2.10. *Minimum bridge cuts with 1 service brake and type A carabiners*

The contribution of type A carabiner failure to the risk of a container dropping is highest and represents 69.8% of the risk. The electrical failure of the motor combined with the failure of another element of the crane (motor encoder, PLC, service brake) contributes 30.2% to the risk.

Assessment for the risk of a container dropping with type B carabiners

With a bridge equipped with a service brake, the annual probability of a container dropping with type B carabiners is 5.8×10^{-3}.

The minimum decisive cuts are shown in Table 2.11.

Rank	Annual probability	Contribution (%)	Event 1	Event 2	Cutting order
1	3.7×10^{-3}	63.5	E031: electrical motor failure	E099: classic PLC failure	2
2	1.1×10^{-3}	19.5	E031: electrical motor failure	E100: encoder failure	2
3	0.9×10^{-3}	16.7	E031: electrical motor failure	E105: service brake failure	2
4	8×10^{-6}	0.1	E049: carabiner failure up 2		1
5	8×10^{-6}	0.1	E048: carabiner failure up 11		1

Table 2.11. *Minimum bridge cuts with 1 service brake and type B carabiners*

The electrical failure of the motor combined with the failure of another element (motor encoder, PLC, service brake) contributed 99.7% to the feared event. The contribution for the failures of the carabiners of type B to the risk of a container dropping is only 0.2%.

The holding difference between the two carabiners does not have a significant effect on the result since the probability of electrical motor failure combined with the failure of another element (motor encoder, PLC, service brake) is far greater than the objective of 5×10^{-5}. The addition of a safety brake was thus evaluated.

Bridge with one service brake and one safety brake

The 5×10^{-5} objective could not be reached with the previous system, so another handling system is studied.

The design including both a service brake and a safety brake is shown in Figure 2.4.

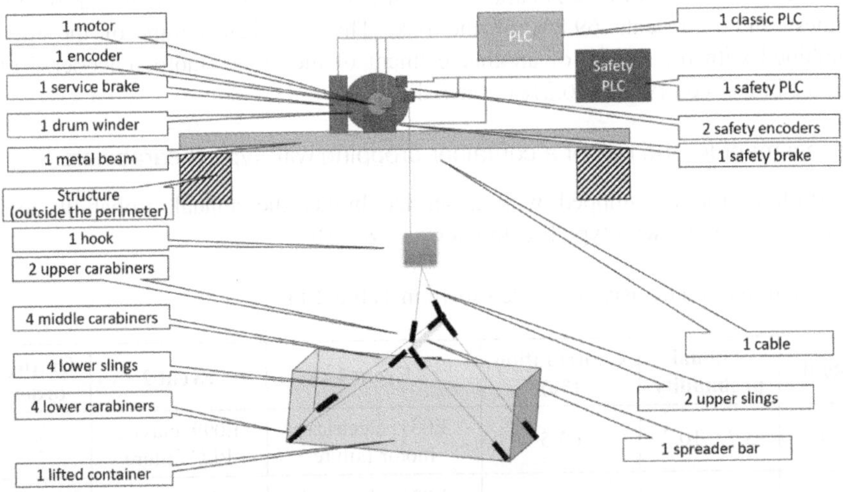

Figure 2.4. *Representation of a handling system with a service brake and a safety brake*

The FMEA was completed with the new components (see Table 2.12).

Description	ID	No.	Technical Functions	Proba of Constant Failure (without unit)	Failure Rate (fail/hour)	Proba of Failure over 1 hour of Operation (without unit)	MTBF (hours)	Source or Justification for Quantification	Comments	MTTR (hours)	Intrinsic Unavailability of the Component Group
Safety Encoder (Magnetic)	C16	1	To acquire drum rotation information architecture, 2 encoders in parallel to improve reliability (comparator in the safety PLC): in this case no optical encoder.	-	1.40E-05	1.40E-05	71,480	NPRD2011 "Rotary Encoder" : 13,99E-6	The encoder makes only one acquisition that allows a computer to deduce the speed and/or acceleration of the drum (and the safety PLC will start, if the speed is judged abnormally high; a safe braking procedure).	8	0.01119%
Safety Encoder (Optical)	C17	1	Acquire information on drum rotation architecture, 2 encoders in parallel to improve the reliability (comparator in the safety PLC). Optical encoder, the redundancy for the magnetic encoder, in order to improve safety and reduce the DCC between the 2 encoders to a maximum.	-	7.13E-05	7.13E-05	14,029	NPRD2011 "Optical Azimuth Encoder": 71.283125E-6	The encoder makes only one acquisition that allows a computer to deduce the speed and/or acceleration of the drum (and the safety PLC will start, if the speed is judged abnormally high; a safe braking procedure).	8	0.05699%
Safety PLC SIL3	C18	1	Ensure the safety braking function on detection of overspeeding.	-	1.00E-07	1.00E-07	10,000,000	Standard IEC61508	The choice of a SIL3 PLC will guarantee control of the risk of random failure on the HW with an MTBF of 10000000 h. SIL3 integrity will also guarantee a level of control of systematic failures, on the HW but also on the SW, which cannot be valued quantatively (except by specifying the respect of SIL3 constraints).	8	0.00008%
Safety Brakes	C19	3	Guarantee fast and efficient safety braking, in all situations requiring it.	-	1.20E-05	1.20E-05	83,542	NPRD2011 "Brake Assembly, Electric", FG	FG = Fixed Ground	20	0.07180%

Description	Description FR	ID	No	Technical Function	Probe (Mission Time)	Failure Mode	Cause	Failure Mode Distribution	λ of Mode (Fail Time)	Annual Effective Probability	System Effect	Severity (Safety/Avail)	Severity FR	Comments
Encoder	Codeur	C16	1	To acquire information about the rotation of the drum.	2.08E-02	Loss	Loss of electrical connection	10%	1.40E-06	2.08E-03	No speed detected by the system although there is one but may be too high.	Safety	Sécu	This defect could be diagnosed by the PLC but not accurately.
Encoder	Codeur	C16	1	To acquire information about the rotation of the drum.	2.08E-02	Absent	Loss of electrical connection	10%	1.40E-06	2.08E-03	No speed detected by the system when there is one but may be too high.	Safety	Sécu	This defect could be diagnosed by the PLC but not accurately.
Encoder	Codeur	C16	1	To acquire information about the rotation of the drum.	2.08E-02	Degraded	Signal degradation	30%	1.12E-05	1.66E-02	Incorrect speed evaluation and risk of overspeeding.	Safety	Sécu	This defect could be diagnosed by the PLC but not accurately.
Classic PLC	Automate classique (PLC)	C6	1	To manage the rotation movement of the winding drum and thus manage the raising or lowering of the load + manage the braking of the load (and thus its maintenance); manage the rotation movements of the drum.	6.77E-02	Loss	Internal failure leading to a lack of brake control	10%	4.67E-06	6.77E-03	No braking control by the PLC and risk of the load dropping.	Safety	Sécu	There is also the risk of losing motion but this is less serious (available).
Classic PLC	Automate classique (PLC)	C6	1	To manage the rotation movement of the winding drum and thus manage the raising or lowering of the load + manage the braking of the load (and thus its maintenance); manage the rotation movements of the drum.	6.77E-02	Lack of braking	Internal failure leading to a lack of brake control	5%	2.34E-06	3.38E-03	No braking control by the PLC and risk of the load dropping.	Safety	Sécu	There is also the risk of losing motion but this is less serious (available).
Classic PLC	Automate classique (PLC)	C6	1	To manage the rotation movement of the winding drum and thus manage the raising or lowering of the load + manage the braking of the load (and thus its maintenance); manage the rotation movements of the drum.	6.77E-02	Erroneous movement	Internal failure leading to untimely motor control	5%	2.34E-06	3.38E-03	Any type of erroneous movement possible (up instead of down, moving instead of braking, etc.) and therefore risk of the load dropping.	Safety	Sécu	Unless there is evidence to the contrary, it must be assumed that there is a non-negigible probability that a PLC will launch an error maneuver. This can be due to the HW and/or SW.
Classic PLC	Automate classique (PLC)	C6	1	To manage the rotation movement of the winding drum and thus manage the raising or lowering of the load + manage the braking of the load (and thus its maintenance); manage the rotation movements of the drum.	6.77E-02	Untimely movement		5%	2.34E-06	3.38E-03	Untimely lifting: risk of being caught and dropping. Untimely lowering: shock on impact with the ground.	Safety	Sécu	Unless there is evidence to the contrary, it must be assumed that there is a non-negigible probability that a PLC will launch an error maneuver. This can be due to the HW and/or SW.
Classic PLC	Automate classique (PLC)	C6	1	To manage the rotation movement of the winding drum and thus manage the raising or lowering of the load + manage the braking of the load (and thus its maintenance); manage the rotation movements of the drum.	6.77E-02	Absence of movement	Internal failure of the PLC	75%	3.50E-05	5.08E-02	Impossible movement, no handling.	Availability	Dispo	The load remains blocked at the bottom.
Drum Reel	Tambour enrouleur	C7	1	To wind the cables to raise the load; to unwind the cables to lower the load.	1.00E-10	Loss	Rupture	100%	No object	1.00E-10	The breakage of the axis or of the connection with the axis of the drum will lead to the untimely unwinding of the drum and thus leads to the load dropping.	Safety	Sécu	

Table 2.12. *FMEA supplement*

Assessment for the risk of a container dropping with type A carabiners

For a crane equipped with a service brake and a safety brake, the annual probability of a container dropping with type A carabiners is 1.4×10^{-2}.

A preview of the decisive cuts is given in Table 2.13.

Rank	Annual probability	Contribution (%)	Event 1	Event 2	Event 3	Cutting order
1	6.7×10^{-3}	49.6	E049: carabiner failure up 2			1
2	6.7×10^{-3}	49.6	E048: carabiner failure up 1			1
3	6.6×10^{-5}	0.5	E031: electrical motor failure	E033: safety brake failure	E099: classic PLC failure	3
4	2×10^{-5}	0.1	E033: safety brake failure	E031: electrical motor failure	E100: encoder failure	3
5	1.7×10^{-5}	0.1	E033: safety brake failure	E031: electrical motor failure	E103: service brake failure	3

Table 2.13. *Minimum bridge cuts with one service brake and one safety brake and type A carabiners*

The failure of type A carabiners is predominant (99.2% of the risk) and does not allow us to see the influence the addition of a safety brake has on the system.

Assessment for the risk of a container dropping with type B carabiners

For a crane equipped with a service brake and a safety brake, the annual probability of a container dropping with type B carabiners is 1.3×10^{-4}.

The decisive cuts are shown in Table 2.14.

Rank	Annual probability	Contribution (%)	Event 1	Event 2	Event 3	Cutting order
1	6.6×10^{-5}	49.7	E031: electrical motor failure	E033: safety brake failure	E099: classic PLC failure	3
2	2×10^{-5}	15.3	E033: safety brake failure	E031: electrical motor failure	E100: encoder failure	3
3	1.7×10^{-5}	13.1	E033: safety brake failure	E031: electrical motor failure	E103: service brake failure	3
4	8×10^{-6}	6	E049: carabiner failure up 2			1
5	8×10^{-6}	6	E048: carabiner failure up 1			1

Table 2.14. *Minimum bridge cuts with one service brake and one safety brake and type B carabiners*

With Type B carabiners, the addition of a safety brake reduces the risk to 1.3×10^{-4}, which is still above the target threshold of 5×10^{-5}. Therefore, the addition of an emergency brake and a second safety brake was evaluated with type B carabiners.

Bridge with one service brake, one emergency brake and two safety brakes

Assessment for the risk of a container dropping with type B carabiners

For a bridge equipped with a service brake, an emergency brake and two safety brakes, the annual probability of a container dropping with type B carabiners is 2.8×10^{-5}, which meets the 5×10^{-5} target.

The minimum cuts with an annual probability of occurrence greater than 10^{-6} are shown in Table 2.15.

Rank	Annual probability	Contribution (%)	Event 1	Event 2	Event 3	Event 4	Cutting order
1	8×10^{-6}	28.1	E049: carabiner failure up 2				1
2	8×10^{-6}	28.1	E048: carabiner failure up 1				1
3	7.8×10^{-6}	27.3	E031: electrical motor failure	E084: magnetic safety encoder failure	E099: classic PLC failure	E104: optical safety encoder failure	4
4	2.4×10^{-6}	8.4	E031: electrical motor failure	E084: magnetic safety encoder failure	E100: encoder failure	E104: optical safety encoder failure	4
5	1.2×10^{-6}	4.1	E031: electrical motor failure	E033: safety brake failure	E099: classic PLC failure	E106: safety brake 2 failure	4

Table 2.15. *Minimum bridge cuts with four brakes and type B carabiners*

If the objective for the risk of a container dropping was less than 10^{-6}, it would have been necessary to review the architecture of the bridge (e.g. a bridge with a closed kinematic chain) and to use specific carabiners. Indeed, the commercial type B carabiner does not satisfy a risk objective of the container dropping that is lower than 10^{-6} (with the probability of failure for a upper carabiner being 8×10^{-6}).

Consideration of a common cause failure

Common cause failures are dependent failures originating from the same direct cause and affecting several redundant components simultaneously. These are therefore important causes at the root of failures in redundant systems.

The general principle is to take into account a single common mode only: for the equipment in service, or stressed at the same time, and as such hold a correlation. A common cause failure must be considered for redundant equipment, identical technologies, identical environments and identical operations.

The model chosen for these common modes is the beta factor method. The ß model is the most widely used single-parameter model to introduce common modes of failure in reliability analyses and to calculate the share of these failures for the failure probability of a system. It associates a ß fraction of the failure rate of a component to a common cause event shared by the other components of the group. This model makes the hypothesis that a common cause causes the systematic failure of all the components of the group. The estimate is pessimistic with respect to the assumption of total independence.

In the following example, we considered a common cause failure for magnetic type safety encoders with a value for the ß factor of 10^{-1}.

The modeling of the failure of magnetic safety encoders without common cause is shown in Figure 2.5.

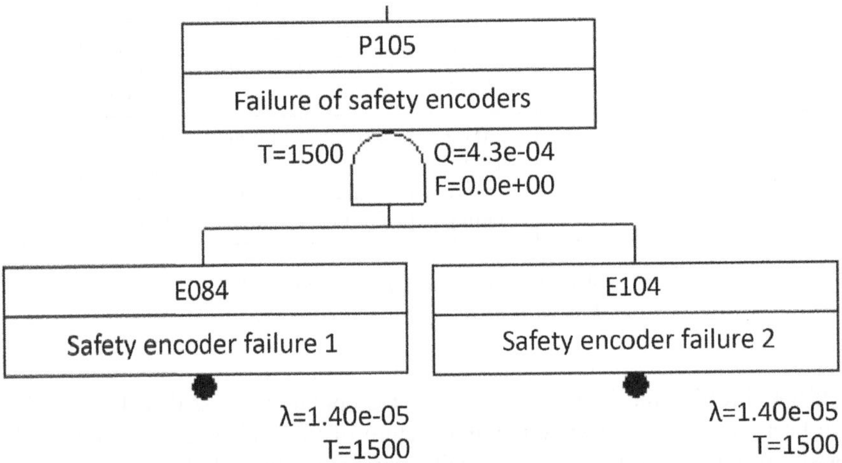

Figure 2.5. *Modeling failure of safety encoders without any common cause*

The modeling for the failure of magnetic safety encoders that takes into account the common cause of failure is shown in Figure 2.6.

The probability of encoder failure decreases from 4.3×10^{-4} without any common cause for a failure (P105 Gate) to 2.5×10^{-3} with a common cause (P119 Gate). The risk of a container dropping increases from 2×10^{-5} without common cause to 3.1×10^{-5} with common cause.

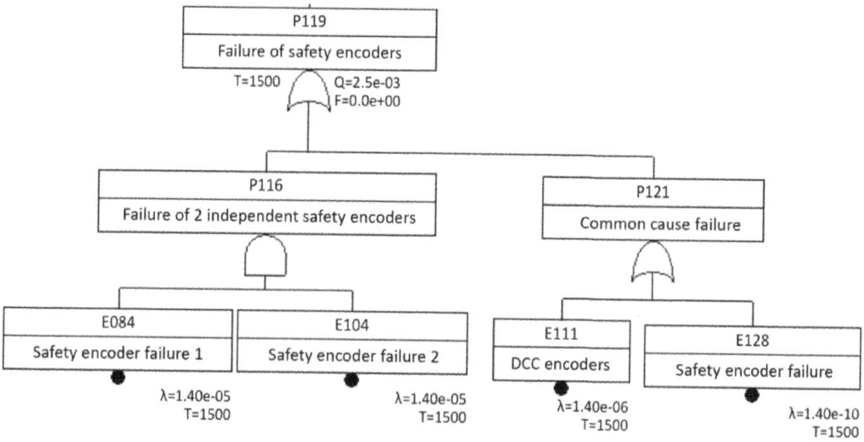

Figure 2.6. *Modeling safety encoder failures with a common cause*

2.4.4.2.5. Modeling the feared "container dropping" event

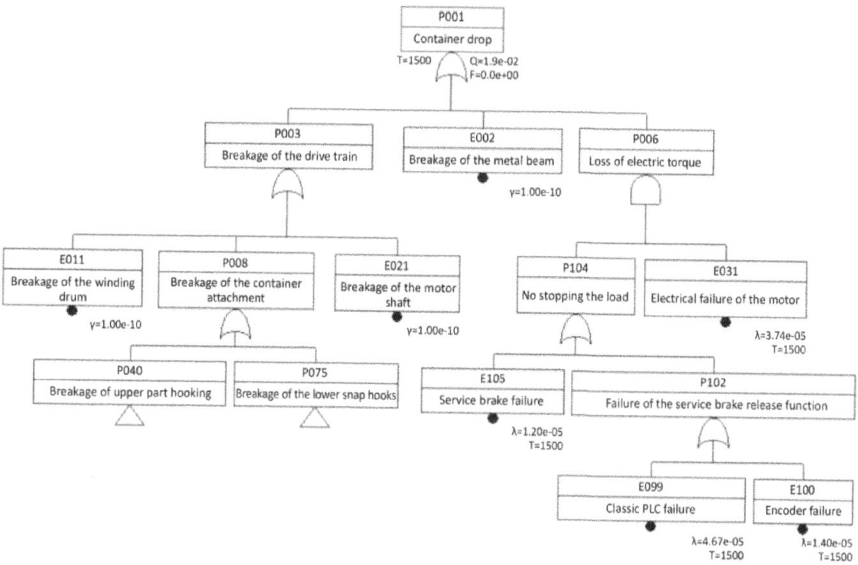

Figure 2.7. *Bridge shaft head with one service brake*

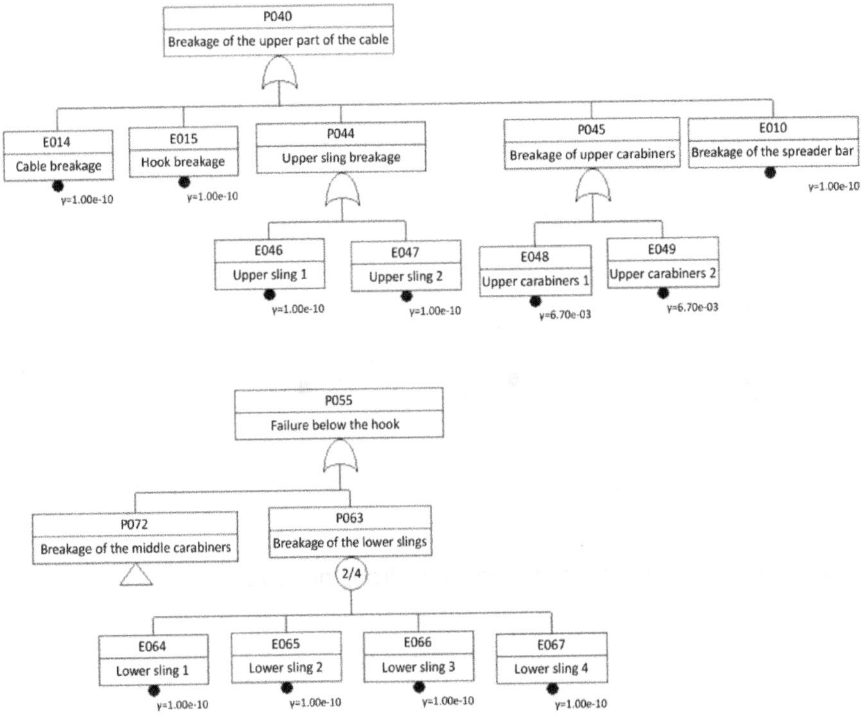

Figure 2.8. *Hanging container (first part) type A carabiner*

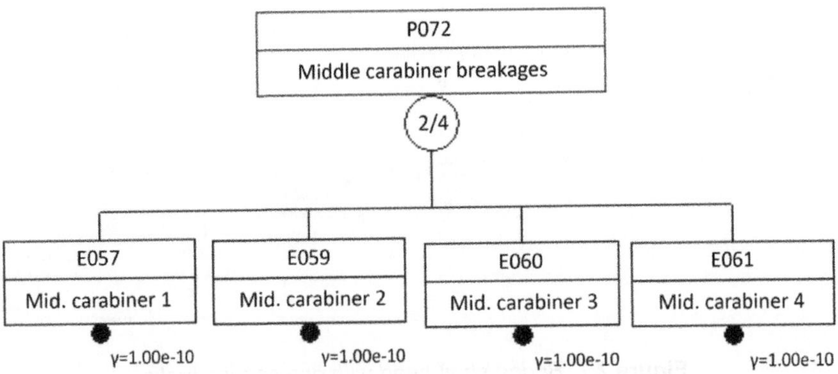

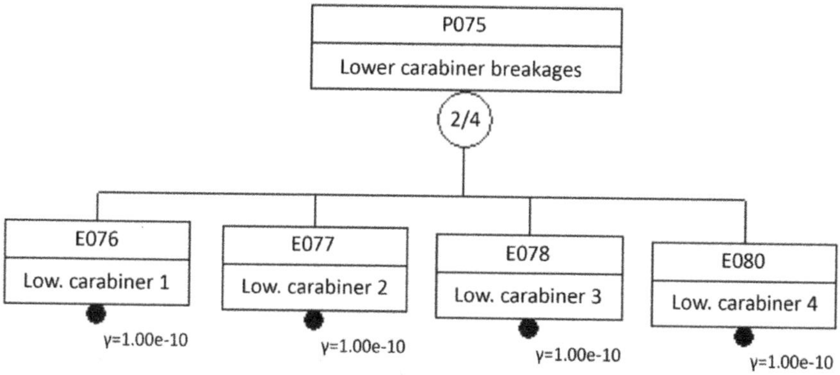

Figure 2.9. *Hanging container (second part)*

The modeling is identical for type B carabiners. Only the value of the failure probabilities for the upper carabiners is modified (8×10^{-6} instead of 6.7×10^{-3}).

Modeling for a bridge with one service brake and one safety brake

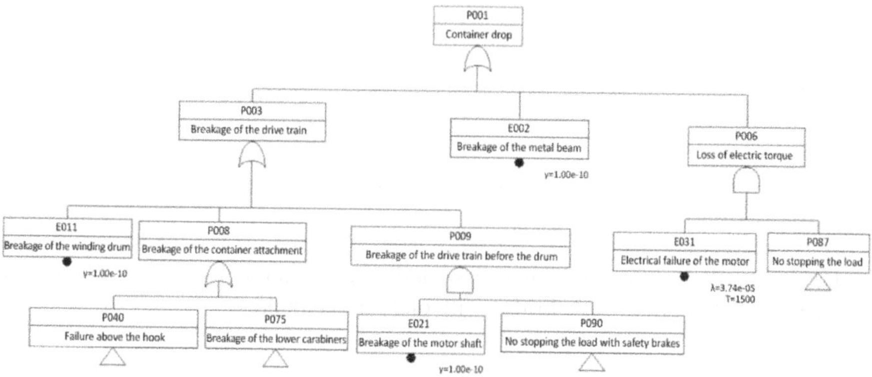

Figure 2.10. *Bridge shaft head with two brakes*

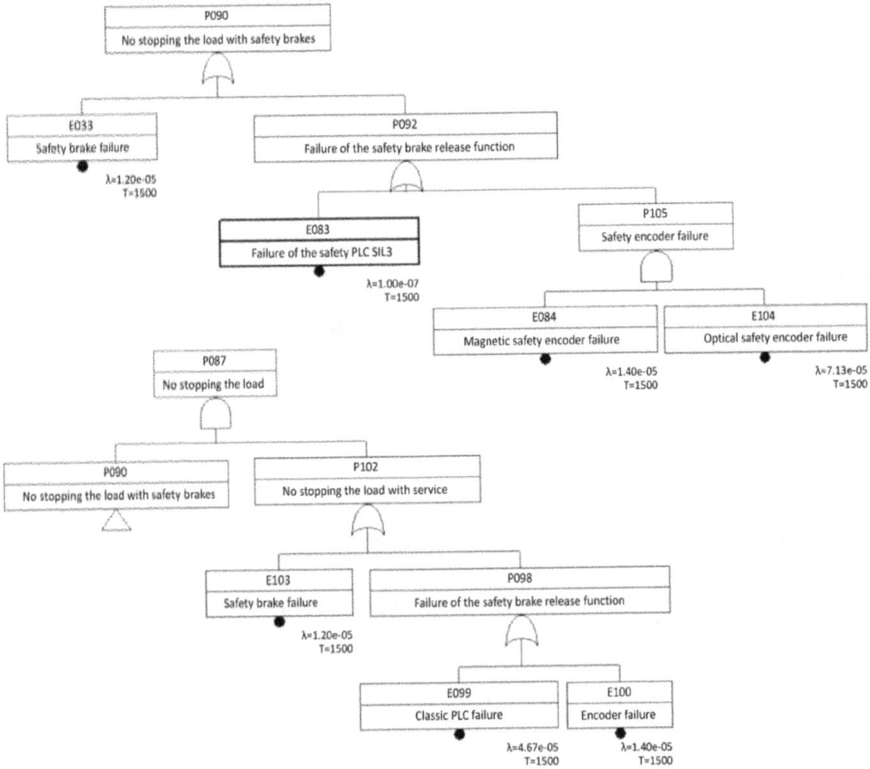

Figure 2.11. *No stop of the bridge load with two brakes*

The other fault trees (hooking first and second parts of the container) are identical to the fault trees for a bridge with one service brake.

Modeling by fault tree analysis a bridge with one service brake, one emergency brake and two safety brakes

The other fault trees (shaft head, first and second container hook-up) are identical to the shafts for a bridge with two brakes.

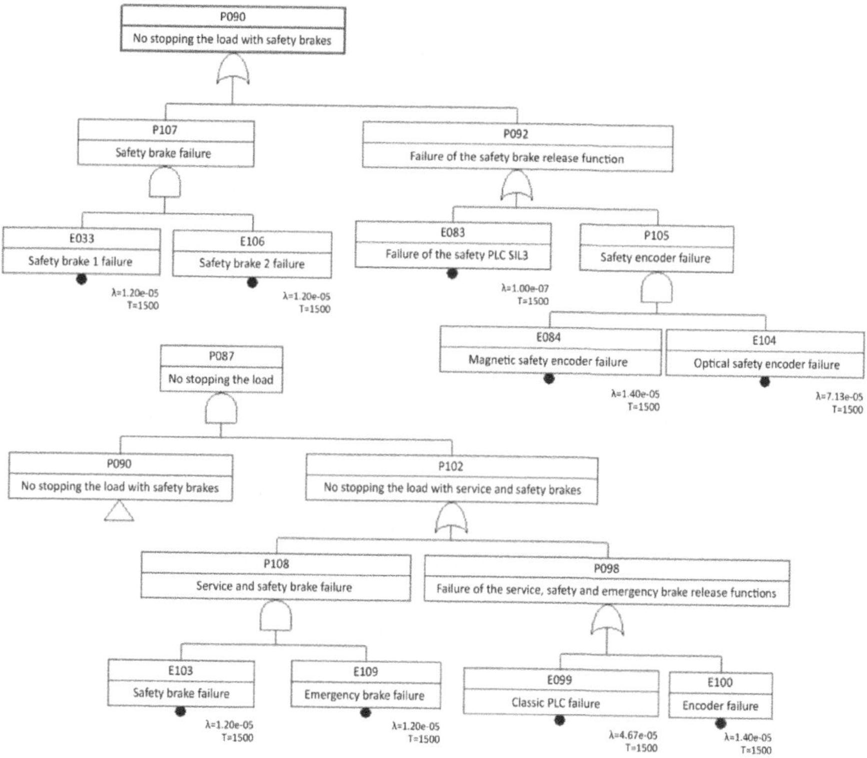

Figure 2.12. *No stopping the bridge load with two brakes*

2.5. Conclusion

2.5.1. *FMECA or fault trees, how to choose?*

This question is regularly asked, as the reliability engineer is subject to financial and time saving pressures, given that the choice cannot simply be imposed arbitrarily.

It is therefore essential to explain the differences and justify their uses.

FMECA:

– is inductive, or "bottom-up", and we go from the cause to the consequence;

– models simple failures and not complex failures;

– allows us to make reliability assessments, MTBF calculations and all the calculations necessary for integrated logistical support (MTTR, operational availability, spare parts calculation, etc.);

– analyzes the model of a system, or a set of functionalities, in a totally exhaustive way (the exhaustiveness to the extent that the model permits); we ask the question for each mode of failure: "what does this mode of failure entail as an effect on the system, what is its severity?"

Fault trees:

– is deductive, or "top-down", and seeks to find the causes of an analyzed consequence;

– through its combinatorial logic, the fault tree is a simple and powerful tool to model complex failures (the conjunction of events);

– it can also be used to make an analysis according to the model of a system, but which then requires the application of a filter by asking the question "what in my system can lead to this event, for example, the risk of a container dropping?";

– /!\ there is a trap here, because this filter has been applied a priori. To be perfectly clear, we need to rely on our own judgment here, or that of a work group, to ensure the list of possible causes is complete. Everything depends on the expertise of this judgment:

- high expertise = low risk of missing cases,

- /!\ low expertise = high risk of missing cases!

Very often, the use of both methods, *FMECA* and *fault trees*, will be essential to guaranteeing an effective and exhaustive modeling that allows for the representation of complex failures:

– FMECA for more exhaustive analyses, and to list all the cause-and-effect links (links without combination) in a system (safety but also others: failures, dissatisfactions, etc.);

– Fault trees for more complex models, with events that are often a combination of failures, and which we will often limit to a few priority events (the most critical ones), such as safety risks for example.

In general, on a system with high criticality, we will "secure" through mechanical and/or electronic "safety mechanisms", and it is thus necessary to have a modeling by fault trees (see Figure 2.13).

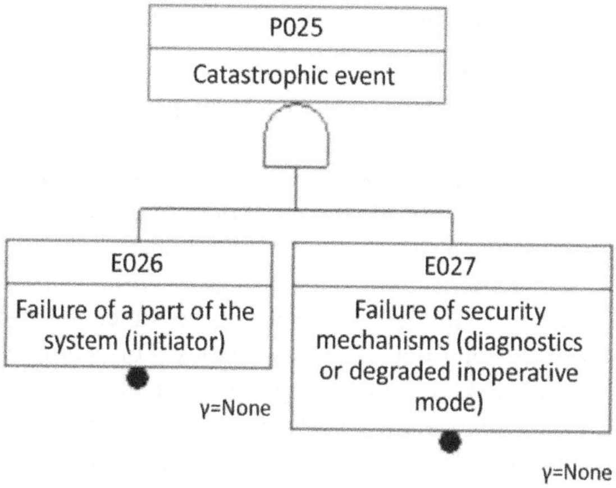

Figure 2.13. *Typical fault tree model of a "secure" system*

2.5.2. *Pitfalls to avoid*

2.5.2.1. *Pitfall No. 1: pooling failures*

In the stress–strength method, strength modeling is generated from failure test results. It is important that these data correspond to the mode of failure for which a reliability objective is being demonstrated. Pooling failures from different modes can distort the analysis.

2.5.2.2. *Pitfall No. 2: probability and failure rate*

For events following an exponential law, the definability is expressed as $D(t) = 1 - e^{-\lambda t}$ and the limited development of e^{-x} enables us to approximate D(t) by λt, when λt is close to 0.

In the case of a reliability calculation over 1 h, the probability of failure is approximated by the failure rate: Pf~λ.

It happens then, by habit of handling low failure rates and from making calculations over 1 h of operation (and this is the case in 90% of the studies!), to make the systematic mental shortcut "Pf = λ" which can be dangerous. Indeed, this approximation is no longer valid if λt is not close to 0, and beyond this numerical error, it creates a confusion between the probability (real without unit in the interval [0 ;1]) and the failure rate (positive real whose unit is the failure/hour), which can lead to multiple illogicalities.

2.5.2.3. *Pitfall No. 3: considering the operating time*

First, in the fault tree modeling it is important to take into account the operating time that corresponds to the handling operations where a safety risk exists.

It is not necessary to multiply a result on 1 h by an annual handling time because the fault trees are made up most of the time, not only of events with time-dependent laws, but also of events with constant failure probabilities.

It is recommended to define, as precisely as possible, the use profile of the studied system before carrying out reliability studies. For example, for a handling machine, it is necessary to know: the annual times of the different phases of use studied (hoisting, translation, steering, low approach, lowering, etc.), as well as the total annual time for when the lifting device is under tension (no-load time, test time), and also the annual number of cycles.

Finally, it is important not to forget to take into account hidden component failures, which go unnoticed. Indeed, a standby component may suffer a failure that is not detected as and when it occurs (as opposed to an active failure). The occurrence of a hidden failure is only visible when the impacted component is under stress, following on from main component failures.

The data needed to calculate the equivalent failure rate for hidden failure components are the time of operation under voltage and the time of storage under voltage between each maintenance service. The equivalent failure rate for the "basic events" components of the "hidden failure" type, that is to say, with a risk of failure when it is under stress, is calculated as follows:

$$\lambda_{equivalent} = \frac{T_f * \lambda + T_S * \frac{\lambda}{R}}{T_T} \qquad [2.13]$$

with:

– λ = failure rate of the component concerned (per hour);

– T_S = storage time between each maintenance;

– $T_T = T_f + T_s$ = total time between each maintenance

$$R = \frac{Operational\ failure\ rate}{Storage\ failure\ rate} \qquad [2.14]$$

In general, R = 20 (mechanical components) or R = 10 (electronic or electro-mechanical components). These values are assumptions used in various fields (including aerospace) and are commonly accepted.

2.5.3. *Note on low reliability targets in innovative systems*

The sometimes-unbridled innovation race leads to the following situation, which is classic but can also be very problematic.

Let us imagine that we are using a component technology whose behavior under certain stresses is not well known, and that the planned system could generate safety risks in the event of component failures.

The reliability tests needed to demonstrate the objective we have set for ourselves will inevitably be very long and very costly. This is unavoidable, and although Bayesian mathematical tricks or acceleration possibilities will produce positive effects, we will not be able to significantly reduce the duration of the tests without having acquired strong experience:

– allowing a good understanding of the phenomena involved, their nature and their characteristics;

– allowing experience feedback data to be capitalized: return on experience (REX).

Therefore, in order to satisfy the strong expectations to reduce the "time-to-market" while keeping a good "risk control", we understand that only the treatment of reliability and the powerful tools of reliability estimation will allow us to define the minimum validations for a desired level of risk.

2.6. General conclusion

The application of a structured RAMS approach, used in a simple but effective way, allows us to attain an overall consistency, and to demonstrate the achievement of qualitative and quantitative objectives.

Therefore, in the fictitious example proposed in this chapter, the choice of a handling system for a container allows us, through a decomposition of the system into a set of components and their functions, to evaluate its reliability (calculation of MTBF), its availability, its maintainability (MTTR) and finally to identify the architecture of the handling system (crane with an open kinematic chain with three brakes and type B carabiners) that will satisfy the target safety objective (risk of container dropping) of 5×10^{-5} per year.

2.7. References

[AAI 14], AAIAG, Automotive Industry Action Group, DRBFM Reference Guide, 2014.

[ASD 15] ASD/AIA S3000L, International procedure specification for Logistic Support Analysis (LSA), 2015.

[BER 17] BERGAMO S., SCHIMMERLING P., TRIBOULET F., et al., Préconisation pour les caractéristiques statistiques de résistance en fatigue, SIA, version: DC-05-01, 2017.

[CLE 14] CLEMENT E., RAUZY A., THOMAS T., Arbre Analyste : un outil d'arbres de défaillances respectant le standard Open-PSA et utilisant le moteur XFTA, IMdR, Institut pour la Maîtrise des Risques, 2014.

[DEL 17] DELAUX D., ECHARD B., LAPEYRE G., et al., Guide d'aide à l'estimation et à la validation de la fiabilité automobile, SIA, version: DC-01-03, 2017.

[LAN 05] LANNOY A., PROCACCIA H., *Evaluation et maitrise du vieillissement industriel*, Lavoisier, Paris, 2005.

[LAN 06] LANNOY A., PROCACCIA H., *Evaluation de la fiabilité prévisionnelle : outil décisionnel pour la conception et le cycle de vie d'un bien industriel*, Lavoisier, Paris, 2006.

[MIL 83] MIL STD 1388, Military standard, logistic support analysis, 1983.

[MIL 03] MIL-HDBK-5J, Metallic materials and elements for aerospace vehicle structures, US Department of Defense, 2003.

[NPR 11] NPRD2011, Nonelectronic Parts Reliability Data 2011, RIAC, 2011.

[NFE 06] NF EN 60812, Techniques d'analyses de la fiabilité du système, Procédure d'analyse des modes de défaillance et de leurs effets (MADE), Edition AFNOR, 2006.

[SOC 13] SOCIETY OF AUTOMOTIVE ENGINEERS, J2886, DRBFM Recommended Practice, 2013.

[YAN 07] YANG G., *Life Cycle Reliability Engineering*, John Wiley & Sons, Hoboken, New Jersey, 2007.

3

Generation of a Failure Model through Probabilistic "*Stress–Strength*" Interaction in a Context of Poor Information

In the industrial field, in particular during the pre-sizing phase of a device, it is rare that the designer has all the data concerning the corresponding mission profile in time. Under these conditions, the implementation of the "*stress–strength*" method usually refers to so-called "*fixed*" information. This information generally concerns the nature of the probability laws on which the interaction process is based, as well as the values of these parameters. The partly arbitrary character of this approach, as well as the resulting choices, lead to forecasts that are, to say the least, distributed: failure probabilities that can vary by a ratio of 10, safety factors by a ratio of 2–3, etc. This has led us to develop and justify an analytical model for interactions based on information theory, whose sufficiently general structure makes it possible to encompass various specific situations.

3.1. Introduction

The most basic version of the "*stress–strength*" method calculates the probability of failure that results from the interaction between two probability laws. In general, if any two laws are explicitly known, the probability of failure can be calculated by digitally solving a convolution integral.

Only a few special cases lead to purely analytical solutions, when the two laws belong to the same statistical family. Among the most classical examples, we can cite the interaction between pairs of homologous laws, normal or lognormal, which allow us to obtain exact analytical solutions [PIE 10].

Chapter written by Lambert PIERRAT.

In other cases, such as the interaction between two *Weibull laws*, we can obtain an approximate analytical solution whose accuracy is sufficient for the needs of industrial practice [PIE 92]. In fact, such situations are not entirely realistic, because generally, the laws derived from statistical estimates deviate from the theoretical laws with which they are assimilated.

Since the resulting deviations are difficult to quantify, the relevance of an exact solution remains partly illusory. The situation is aggravated when, at the preliminary design stage, there is only poor knowledge of the mission profile available.

In a pre-dimensioning context, the designer has to compensate for a lack of information by using so-called "*lump sum*" values, supposedly based on specific "*user experience*" from the application domain that concerns the device. Such an approach involves a certain amount of arbitrariness because generally, the "*lump sum*" values are neither clearly nor rigorously justified.

3.2. Aims and objectives

In the industrial domain, the extreme diversity of laws likely to intervene in a "*stress–strength*" interaction led Wirsching [WIR 76] to seek a sufficiently general model, capable of covering a large number of cases.

With this perspective, he proposed a model based on the interaction between an exponential law (stress) and a power law (resistance). Although it meets the objective he had set, the use of this model faces several different kinds of limitations:

– its mathematical structure, fixed a priori by statistical laws, is not based on physical arguments;

– the power law must be truncated by upper values, which requires choosing a somewhat arbitrary level whose influence is difficult to apprehend;

– in all cases, obtaining the failure probability requires numerical integration;

– it is not possible to invert this model, which does not enable us to explicitly calculate the safety factor corresponding to the failure probability.

In order to overcome these limitations, the objective of this chapter is to develop and justify a sufficiently general interaction model, allowing us to analytically estimate a probability of failure and a safety factor, in the context of minimal information. It is thus possible to justify the determination of "*lumped*" values whose consideration will lead us to increase the results based on the interaction of various types of different laws.

The article successively addresses the following points:

– in the first part, the choice of a stress law (exponential) and a resistance law (normal) is justified by physical and informational considerations (i.e. information entropy);

– in the second part, the analytical solution for the failure probability resulting from the interaction between these two laws is developed;

– in the third part, at the cost of a legal approximation, we deduce from this solution the safety factor that allows the designer to apprehend, and even to optimize, the sizing of the device;

– in the last part, we expose a quantitative validation procedure as well as some applied examples.

3.3. Choosing types of legislation

Choosing the types of probability laws likely to represent a constraint and a resistance must simultaneously meet several different criteria.

On the one hand, their general form must be compatible with the reality of the physical process concerned.

On the other hand, they must lead to a sufficiently simple and general analytical formulation, in the context of minimal information.

3.3.1. *Principle of maximum entropy*

The quantity of information contained in a continuous probability law is directly related to the infinite statistical moments.

However, when their formulation is limited in number (mean, variance, skewness, kurtosis), they do not condense all the information.

A more relevant measure is that of the differential entropy initially formulated by Shannon [SHA 48], which is expressed as a function of the probability density of a continuous law through the integral form:

$$H_f = -\int_{\Omega} f(x) \cdot Ln[f(x)] \cdot dx \qquad [3.1]$$

Later, the principle of maximum entropy formulated by Jaynes [JAY 57] made it possible to define, in a unique way, a continuous probability law whose maximum entropy takes into account various conditions that were imposed a priori.

These are related to the amount of information available, generally determined by the existence of statistical moments (theoretical or empirical), as well as by the support imposed on the random variable concerned. These specific conditions will be specified later in order to obtain laws compatible with physical reality.

The implementation of the maximum entropy principle relies on the calculation of variations, more precisely by *Lagrange multipliers* [FOX 87] that determine the structure of the law and its parameters.

3.3.2. *The strength distribution*

The choice of a normal distribution is simultaneously justified by two types of consideration, one physical in nature and the other informational in nature.

3.3.2.1. *Physical arguments*

In essence, the designer has relatively precise information on the resistance of a device, because they have control over its physical sizing.

This is the case in particular for actual materials, whose quality level results in a relatively low dispersion of their mechanical characteristics. In most cases, if we are concerned that failure mechanisms exceed a limit value of resistance, such coefficients of variation rarely exceed 10% [AME 61, HAU 65].

This makes it possible to assimilate the random resistance of a mechanical component to a normal law, denoted $N(\mu_r,\sigma_r)$, having as its probability density:

$$f_r(x,\mu_r,\sigma_r)=\left(1/\sigma_r\cdot\sqrt{2\pi}\right)\cdot\exp-(1/2)\cdot[(x-\mu_r)/\sigma_r]^2 \qquad [3.2]$$

This law being defined by the unlimited support $\Omega\in\left]-\infty,+\infty\right[$ must be truncated at the origin in order to respect the positive character of the physical variables.

Its coefficient of variation is thus usually limited to a specific value, such as:

$$\gamma_r=\sigma_r/\mu_r\leq 1/3 \qquad [3.3]$$

This maximum value determines a truncation of the lower tail, corresponding to 3 standard deviations, with a practically negligible probability of exceeding this, that is, 0.135%.

3.3.2.2. *Informative arguments*

Using information theory, the choice of this type of law is justified through the application of the principle of maximum entropy, which can be summarized as follows:

If a random variable, defined on an infinite support $\Omega \in \left]-\infty,+\infty\right[$ is only determined by a mean value (μ_r) and a variance (σ_r^2), it obeys a normal law $N(\mu_r, \sigma_r)$.

The minimal amount of information linked to the knowledge of the two parameters is necessary and sufficient, with the knowledge that the existence of a variance reflects the uncertainty induced by the technological dispersion.

3.3.3. *The law of constraint*

The choice of an exponential law is justified jointly by two types of considerations, one statistical in nature and the other informational in nature.

3.3.3.1. *Statistical arguments*

The probabilistic interaction process that leads to failure by exceeding the resistance is influenced essentially by the severity of the extreme stresses. Statistically, this justifies the use of laws with a probability density whose upper tail slowly decreases, characterizing an inverse relationship between the severity of the stresses and their probability of occurrence.

From this point of view, the lognormal law is the one with the heaviest tail, but other laws have a similar appearance: exponential, *Pareto*, power laws, etc. [SCH 84]

Among these, the preferential choice of an exponential law can be conceived in a stochastic context, its memoryless property, is linked to the purely random event flow generated by a *Poisson process* [PAP 84].

3.3.3.2. *Informative arguments*

As with the law of resistance, this choice is confirmed by the application of the principle of maximum entropy, which leads to the following property:

If a random variable, defined on a positive support $\Omega \in [0,+\infty[$ is determined only by a mean value (η_c), it obeys an exponential distribution function:

$$F_c(x,\eta_c) = 1 - \exp-(x/\eta_c) \qquad [3.4]$$

The physical reality corresponds well to positive variables; it only has an average value for a minimal quantity of information.

Due to its memoryless property, it has a unitary variation of the coefficient.

3.3.4. *Relationship between the two laws*

It is possible to demonstrate a mathematical relationship between the exponential law and the normal law through several successive statistical transformations, involving the *Rayleigh law*.

3.3.4.1. *Transition from the normal distribution to the* Rayleigh law *distribution*

In a two-dimensional space, consider a vector whose two orthogonal components (U,V) are random variables distributed according to two standardized normal laws $N(0,\sigma)$, mutually independent, and of the same standard deviation (σ).

Next, the amplitude of the vector is defined by the relationship:

$$X = \sqrt{U^2 + V^2} \qquad [3.5]$$

Its distribution obeys a Rayleigh law with probability density:

$$f(x,\sigma) = \left(x/\sigma^2\right) \cdot \exp-\left(x/\sigma\sqrt{2}\right)^2 \qquad [3.6]$$

This law, initially developed by Rayleigh [RAY 96], has subsequently been the subject of numerous and diverse applications in the field of physics.

3.3.4.2. *Transition from* Rayleigh's law *to the exponential law*

We apply to the random variable distributed according to a *Rayleigh distribution* [3.6] a quadratic transformation such that:

$$y = x^2 \;\; versus \;\; x = \sqrt{y} \qquad [3.7]$$

Next, the random variable is thus transformed [PAP 84] by applying an exponential law:

$$f(y,\sigma)=\left(1/2\sigma^2\right)\cdot\exp-\left(y/2\sigma^2\right) \quad [3.8]$$

3.3.4.3. *Energy considerations*

Let us consider, as an illustration, the simple case of a linear mechanical system of the second order that has been excited in a vibratory regime that is low in amplitude.

The Lagrangian approach to its behavior can be summarized in two functions (**H**, **L**), combining the kinetic energy (**T**) and the potential energy (**U**), both quadratic functions of the generalized coordinates [WIT 37]. These are, on the one hand, the Lagrangian $L \equiv (T-U)$, and, on the other hand, the Hamiltonian $H \equiv (T+U)$, which are, in particular, the basis of the *Rayleigh–Ritz method*, widely used in the field of nonlinear mechanics [RIT 09].

This aims to demonstrate that the quadratic transformation has a general physical character related to the energy of a dynamical system, which can be compared with the previous statistical transformation.

3.4. Probability of failure

Starting from a general formulation, the introduction of a change in variable leads to a parametric expression adapted to the "*stress–strength*" context. This enables us to obtain the exact analytical expression for the probability of failure.

3.4.1. *Formulation*

The probability of failure, which results from the probabilistic interaction between the two laws, can be written as two equivalent convolution integrals.

Depending on the laws considered, the choice of either facilitates the obtainment of an analytical solution of the integrals in the following form:

$$P_f = 1-\int_0^\infty F_c(x)\cdot f_r(x)\cdot dx \equiv \int_0^\infty F_r(x)\cdot f_c(x)\cdot dx \quad [3.9]$$

In our case, the stress is distributed according to an exponential distribution with the distribution function:

$$F_c(x,\eta_c)=1-\exp-(x/\eta_c) \tag{3.10}$$

The resistance being distributed according to a normal distribution with a probability density of:

$$f_r(x,\mu_r,\sigma_r)=\left(1/\sigma_r\cdot\sqrt{2\pi}\right)\cdot\exp-(1/2)\cdot\left[(x-\mu_r)/\sigma_r\right]^2 \tag{3.11}$$

By introducing [3.10] and [3.11] into [3.9], we get the solution of the definite integral:

$$P_f=1-\int_0^{\infty}\left[1-\exp-\left(\frac{x}{\eta_c}\right)\right]\cdot\left(\frac{1}{\sigma_r\cdot\sqrt{2\pi}}\right)\cdot\exp\left[-(1/2)\cdot\left(\frac{x-\mu_r}{\sigma_r}\right)^2\right]\cdot dx \tag{3.12}$$

In this integral, there appears two terms:

– one is the complement for the distribution function of the standardized normal law $\Phi(-\mu_r/\sigma_r)$, relative to the law of resistance;

– the other is an exponential function whose argument combines the parameters of the resistance and stress laws.

3.4.2. *Analytical solution*

By introducing into the argument this exponential function, the change of variable is given as:

$$z=\left[\frac{x-\left(\mu_r-\sigma_r^2/\eta_c\right)}{\sigma_r}\right] \tag{3.13}$$

and by changing the lower bound of $x=0$ at $z_0=\left[-\frac{\left(\mu_r-\sigma_r^2/\eta_c\right)}{\sigma_r}\right]$,

the integral defined finally reduces to:

$$P_f=\left[1-\Phi\left(\frac{\mu_r}{\sigma_r}\right)\right]+\Phi\left[\frac{\left(\mu_r-\sigma_r^2/\eta_c\right)}{\sigma_r}\right]\cdot\exp-(1/2)\cdot\left[(2\mu_r/\eta_c)-(\sigma_r/\eta_c)^2\right] \tag{3.14}$$

3.4.3. *Parametric expression*

In order to bring solution [3.14] into the usual context of the "*stress–strength*" method, it is necessary to introduce two new dimensionless variables.

It is, on the one hand, the coefficient of variation for the strength $(\gamma_r = \sigma_r / \mu_r \leq 1/3)$, knowing that that of the stress is intrinsically unitary $(\gamma_c \equiv 1)$, by definition of the exponential law.

On the other hand, it is the safety factor which is identified by the ratio between the averages for the resistance and the stress, that is, $\rho = (\mu_r / \eta_c) > 1$.

It allows the exact solution [4.6] to be written in the more explicit form:

$$P_f = \left[1 - \Phi\left(\frac{1}{\gamma_r}\right)\right] + \Phi\left[(1/\gamma_r) \cdot \left(1 - \gamma_r^2 \cdot \rho\right)\right] \cdot \exp - \rho \cdot \left[1 - \left(\gamma_r^2 \cdot \rho/2\right)\right] \qquad [3.15]$$

This expression for the probability of failure as a function of the parameters (γ_r, ρ), can be used to check the dimensioning of an existing or device still in development.

Its use requires the introduction of the distribution function for the standardized normal distribution:

$$\Phi(X) = \left(\frac{1}{\sqrt{2\pi}}\right) \cdot \int_0^X \exp - \left(\frac{u}{\sqrt{2}}\right)^2 \cdot du \qquad [3.16]$$

In fact, as we shall see below, obtaining a simpler expression leads to the disappearance of the integral function [4.8], thus avoiding the need to consult tables [SPI 68], or to determine the required numerical value.

3.5. Safety factor

There is a close relationship [MIS 70] between the probability of failure and the safety factor. As such, it is necessary to simultaneously work on these two parameters in order to achieve a reliable design.

From this point of view, it is interesting to have an analytical expression for the safety factor, obtained by inverting the expression for the probability of failure.

3.5.1. *Simplified expressions*

The expression [3.15] cannot be inverted analytically, which makes it impossible to estimate the safety factor as a function of an assigned probability, and thus to predict the sizing of a device during design.

However, this problem can be solved approximately by eliminating in [3.15] some terms whose value is either negligible or close to one.

With this in mind, the maximum value assigned to the coefficient of variation for the resistance, the first term, increased by ($Max\left[1-\Phi(3)\right] < 1.35 \cdot 10^{-3}$), is practically negligible.

As for the multiplier coefficient of the exponential term, it is increased by a single unit, with a negligible error if ($\gamma_r^2 \cdot \rho \to 0$), but which increases significantly as soon as ($\gamma_r^2 \cdot \rho \to 1$).

The dependency relations between the trio of variables (P_f, γ_r, ρ) leads to a relative compensation between these two approximations. We derive the following approximate expression:

$$P_f \approx \exp - \rho \cdot \left\{1 - \left(\gamma_r^2 \cdot \rho/2\right)\right\} \qquad [3.17]$$

Given the quadratic form of the safety factor, this expression can be inverted, resulting in the approximate explicit solution:

$$\rho \approx \left(1/\gamma_r^2\right) \cdot \left\{1 - \sqrt{1 - 2\gamma_r^2 \cdot Ln\left(1/P_f\right)}\right\} \qquad [3.18]$$

3.5.2. *Validity limits*

The validity of these two expressions requires the limitation of the combined ranges of variation for the three parameters (P_f, γ_r, ρ), in order to contain approximation errors to practically admissible values in an industrial context of pre-dimensioning.

3.5.2.1. *Theoretical validity intervals*

If the coefficient for the variation of resistance $(\gamma_r \to 0)$, the probability of failure and the safety factor are related by a very simple relationship and bounded by the following values:

$$P_f > \exp - \rho \text{ versus } \rho > Ln(1/P_f) \quad [3.19]$$

In this limited case, the interaction process is reduced to a probability of failure corresponding to a quantile of the exponential law equal to the value of the safety factor.

The existence of a finite probability of failure implies that the safety factor is a positive dimensionless quantity, that is, $(\rho > 0)$.

With respect to [3.18], this leads to a relation of dependence between the variables (P_f, γ_r) allowing us to define an upper limit for the coefficient of variation (γ_r^u) according to the probability of failure (P_f), such as:

$$\gamma_r^u = \sqrt{1/2 \cdot Ln(1/P_f)} \quad [3.20]$$

As for the safety factor, this is framed by upper (ρ^u) and lower (ρ^l) values as follows:

$$\rho^l = \rho^u/2 \leq \rho \leq \rho^u = (1/\gamma^u)^2 = 2 \cdot Ln(1/P_f) \quad [3.21]$$

3.5.2.2. *Practical validity intervals*

Table 3.1 groups all of the previous results according to a series of values for the probability of failure.

P_f	$5 \cdot 10^{-2}$	$3 \cdot 10^{-2}$	10^{-2}	$5 \cdot 10^{-3}$	$3 \cdot 10^{-3}$	10^{-3}	3.10^{-4}	10^{-4}	10^{-5}
γ_r^u	0.409	0.378	0.329	0.307	0.294	0.269	0.248	0.233	0.208
ρ^u	5.99	7.01	9.21	10.60	11.62	13.82	16.22	18.42	23.03
ρ^l	3.00	3.51	4.61	5.30	5.81	6.91	8.11	9.21	11.51

Table 3.1. *Parametric validity intervals*

The analysis of these theoretical results leads to the definition of more restricted intervals of use, taking into account certain orders of magnitude assigned to the parameters.

A first limitation, imposed on the coefficient of variation by the truncation at the origin of the normal law of resistance, authorizes the use of this maximum value for probabilities of failure higher than 1%.

In industrial practice, this condition is acceptable, because apart for safety devices, probabilities for failure of 1% and 5% are commonly used.

A second limitation can be imposed on the safety factor, because the upper tail for the exponential law of stress logically results in higher values the lower the probability of failure. In the most common applications, it is realistic to limit the torque defined by [3.19] to $\{P_f \geq 10^{-3} \Leftrightarrow \rho^l \leq 6.91\}$, corresponding to values for the coefficient of variation $(\gamma_r^u \leq 26.9\%)$.

These limitations logically result from the majorized character of the model, in the sense that it aims to encompass the results of other more specific and less constraining types of interaction.

3.6. Validation and applications

First, a quantitative validation procedure is reviewed by comparing the results of the "*exponential/normal*" distribution for the two different models.

Next, we present three applied cases of the model related to the fields of instrumentation, civil engineering and electrical engineering.

3.6.1. *Comparative analyses*

In order to illustrate quantitatively the previous considerations, we consider a reference case corresponding to the proposed model, defined by a normal resistance law $(\gamma_r = 19\%)$ and by an exponential stress law $(\eta_c = 10)$.

The failure probabilities are calculated from expression [3.17], according to the safety factor that varies within the interval $\{\rho \in [3-8]\}$.

We compare the results of this reference model with those of two other models of "*stress–strength*" interactions.

The first is an "*exponential/gamma*" model for which an exact analytical solution can be obtained, the resistance law being assimilated to the normal law of reference.

The second is a "*log-normal/log-normal*" model for which an exact analytical solution can also be obtained. In this particular case, the stress law, by assimilation to the exponential law, is affected by a unitary variation ratio, whereas the resistance law is assimilated to the normal law of reference.

3.6.1.1. *"Exponential/gamma" interaction*

The probability density of the gamma distribution is written [PAP 84] as:

$$f_r^G(x,\eta_G,\beta_G)=[1/\eta_G\cdot\Gamma(\beta_G)]\cdot(x/\eta_G)^{\beta_G-1}\cdot\exp-(x/\eta_G) \qquad [3.22]$$

Its shape parameter (β_G) depends only on the imposed coefficient of variation, which is:

$$\beta_r^G>1=(1/\gamma_r)^2=(1/0.19)^2\approx 27.7 \qquad [3.23]$$

Its scale parameter is included in the safety factor as:

$$\eta_G=(\rho\cdot\eta_c/\beta_G) \qquad [3.24]$$

Compared to the normal law of reference, whose asymmetry is zero by definition, the coefficient of asymmetry for the gamma distribution is positive, equal to double the coefficient of variation $(Sk^G=0.38)$, which translates the existence of a more slowly decreasing upper tail.

The convolution integral that governs the probabilistic interaction between an exponential stress law and a gamma strength law has an exact solution:

$$(1/P_f^G)=(1+\rho)^{\beta_G} \qquad [3.25]$$

3.6.1.2. *"Log-normal/log-normal" interaction*

The exponential transformation of a normal random variable $\mathrm{N}(\mu,\sigma)$ is distributed according to a log-normal distribution, denoted $L\mathrm{N}(\mu,\sigma)$, having a probability density [AIT 57] of:

$$f^{LN}(x,\mu,\sigma)=\left[1/\sigma\cdot x\sqrt{2\pi}\right]\cdot\exp-\left[Ln(x)-\mu/\sigma\cdot\sqrt{2}\right]^2 \qquad [3.26]$$

An explicit determination of the parameters (μ,σ) relating to the stress and resistance laws is not necessary, since the probability of failure resulting from their interaction is identified with the quantile (β_{LN}) for the standardized normal distribution, whose values are tabulated [SPI 68] as:

$$Pf^{LN}=\Phi(-\beta_{LN}) \qquad [3.27]$$

This quantile (reliability index) depends only on the safety factor (ρ) and the coefficients of variation relating to the stress $(\gamma_c=100\%)$ and strength $(\gamma_r=19\%)$:

$$\beta_{LN}=Ln\left[(\rho)\cdot\sqrt{2/(1+\gamma_r^2)}\right]\Big/\sqrt{Ln\left[(2)\cdot(1+\gamma_r)^2\right]} \qquad [3.28]$$

3.6.1.3. *Results and interpretation*

Table 3.2 groups the previous results, in which the failure probabilities are calculated as a function of the safety factor.

Those from the proposed model are calculated from expression [3.17] and are taken as references for the comparative analysis.

(ρ)	3	4	5	6	7	8
$P_f^G(\%)$	5.79%	2.38%	1.01%	0.44%	0.19%	0.089%
$P_f(\%)$	***5.86%***	***2.44%***	***1.06%***	***0.47%***	***0.22%***	***0.106%***
(β_{LN})	1.672	2.009	2.271	2.484	2.665	2.821
$P_f^{LN}(\%)$	4.75%	2.21%	1.16%	0.66%	0.51%	0.24%

Table 3.2. *Comparative analyses. For a color version of this table, see: http://www.iste.co.uk/elhami/applied1.zip*

It can be seen that over the whole variation interval of the safety factor considered, the results of the *"exponential/normal"* distribution model systematically exceed those of the *"exponential/gamma"* model, that is, $\left\{P_f > P_f^G\right\}$. This is consistent with the objective of the proposed model.

On the other hand, with the exception of safety factors $(\rho \leq 4.5)$, the results of the *"log-normal/log-normal"* model are greater than those of the reference model, the more so as the safety factor increases.

This tendency is explained by the fact that the exponential stress law has been assimilated to a lognormal law with a unit coefficient of variation. However, among all the unimodal laws, the probability density of the lognormal law is the one with the heaviest upper tail (i.e. the slowest decay). For the same unitary coefficient of variation, the skewness coefficient of the lognormal distribution is much higher than that of the exponential distribution $\left\{Sk^{LN} = 4 > Sk^{EX} = 2\right\}$.

This tendency for the *"log-normal/log-normal"* interaction to increase does not call into question the validity of the *"exponential/normal"* interaction model, because the foundations of the latter are the most relevant.

Indeed, arguments based on information theory prevail over arguments based solely on statistical transformations.

3.6.2. *Applications*

The proposed model has many potential applications, due to the quadratic transformation of *Rayleigh's law.*

As an indication, this law often arises in the domain of aerodynamics, such as the distribution of the wind speed [PIE 08], in the domain of electromagnetic compatibility, such as the distribution of an unbalance criterion for a three-phase electrical network [PIE 93], in the domain of telecommunications and signal processing, such as the maximal distribution of a narrow-band vibratory process [DAV 58], etc.

We present hereafter three applied cases related to different technical fields: the reliability of a measuring instrument; the reliability of a civil engineering structure and the reliability of an electrical cable.

3.6.2.1. *Reliability of a measuring instrument*

In [BUR 78], the author presents the case of a measuring instrument subjected to mechanical forces to which it has been subjected during the course of its transport.

The objective is to determine the resistance of the device subjected to $(n = 1{,}235)$ stresses by a random force whose amplitude is characterized by a *Weibull law* [WEI 51], denoted as: $W(\eta = 2.71N, \beta = 2.32)$, so that its reliability reaches $(1 - P_f = 95\%)$.

Without considering the additional influence of the uncertainty affecting the parameters of the *Weibull distribution* (estimated from a limited number of observations), he obtains the exact solution $F_r(n = 1235) \approx 7.34N$.

Reduced to the mean value of the *Weibull distribution* $(\mu_w \approx 2.401N)$, it corresponds to a safety factor of $(\rho \approx 3.057)$.

Since this result was obtained without explicitly considering the stress law, we propose here to show that its assimilation to an exponential law leads to a satisfactory prediction.

The number of stresses being sufficiently high, we can consider that the resistance law tends asymptotically and with a satisfactory accuracy toward an extremal *Gumbel law* [GUM 58].

The analytical determination for the two parameters of this law takes into account the two parameters of the parent Weibull law and the number of stresses, which is noted as: $G_c(u_n \approx 6.315N; \sigma_n \approx 0.382)$.

Equating the upper tail of the probability density of this extremal distribution to an exponential distribution, leads to the following approximate identity:

$$P_f \approx \exp\left[-\frac{(\rho \cdot \mu_w) - u_n}{\sigma_n}\right] \qquad [3.29]$$

It enables us to explicitly estimate the safety factor as follows:

$$\rho \approx (1/\mu_w) \cdot \left[u_n + \sigma_n \cdot Ln(1/P_f)\right] \approx 3.107 \qquad [3.30]$$

Compared to the exact safety factor, the increase is less than (+ 1.64%), which confirms the validity of the exponential character of the tail of the extremal law.

3.6.2.2. *Reliability of a civil engineering structure*

In [LIN 93], the author applies the simplified version of the *"stress–strength"* method for a proposed technical economic study on the predictive reliability of a dam located within an earthquake zone.

The stress is the horizontal acceleration generated by the earthquake over a period of 5 years, whose amplitude varies randomly according to an exponential law.

This stress acts on the structure whose resistance, itself random, is distributed according to a normal law for the coefficient of variation $(\gamma_r = 25\%)$.

The failure probability corresponding to a safety factor $(\rho = 6)$ is calculated numerically from the convolution integral.

The accuracy of the explicit solution from the simplified expression [3.17] can therefore be evaluated, compared to the exact solution from expression [3.15].

The relative error affecting the probability of failure calculated from [3.17], $P_f \approx 0.76351\%$, is less than 0.64%, and is thus practically negligible.

It is also observed that the pair $(\rho = 6, P_f \approx 0.76\%)$ is well within the bounded intervals indicated in Table 3.1.

3.6.2.3. *Reliability of an electrical cable*

Underground cables are components that carry energy within an electrical network. Their reliability is essential given the long-life expectancy for this type of installation.

For reasons of dielectric strength, these are made from insulating materials that are susceptible to thermal degradation. Their technical characteristics, technology and operating conditions determine their surface temperature, which limits their current-carrying capacity [IEC 82].

In the context of the *"stress–strength"* method, it is a random threshold, weakly dispersed, similar to a resistance, distributed approximately according to a normal distribution.

The stress is itself a random variable, a function of a combination of various parameters. Without going into detail, its determination must take into account *a minima* the following random variables:

– functional variables: ambient temperature profile and load profile that causes *Joule effect* losses, which lead to the heating of the active parts of the cable;

– structural variables: thermal resistance of the soil, depending on its nature, its humidity level and the depth the cable is buried at.

In a steady state, the surface temperature results from a thermal balance that takes into account all these parameters. The thermal field is modeled using a finite element method [ZIE 65].

Among all the parameters involved, a deterministic sensitivity study has shown that the variability of three parameters essentially determines the variability of the surface temperature [IND 84]: the load profile (which determines the losses, hence the conductor temperature), the thermal resistance of the soil and the ambient temperature.

The analysis of the results for a probabilistic study carried out by Anders et al. [AND 88] makes it possible to assimilate with an acceptable level of precision the tail of the probability density for the surface temperature (i.e. the stress) to an exponential law.

Given the monthly average value of the ambient temperature at the location being considered $\left(\bar{\theta}_a \approx 21°\text{C}\right)$, the surface temperature is equal to $\left(\bar{\theta}_c\right) \approx 31°\text{C} > \bar{\theta}_a$.

When the cable is stressed by a temporary overload (i.e. the network in a situation of emergency), the maximum admissible temperature at the surface of the cable is a random variable of normal law $\text{N}(\mu_r = 120°C, \gamma_r = 10\%)$.

Under these conditions, the safety factor $(\rho \approx 3.871)$ enables us to determine the probability of failure from expression [5.1], that is, $\left(P_f \approx 2.25\%\right)$.

In accordance with the property of the *"exponential/normal"* model, this value increases the value $\left(P_f \approx 2.0\%\right)$ obtained through numerical integration from the empirical law, taking into account all the influencing factors.

This has a relatively high probability of failure that seems to contradict the reliability objective of a cable subjected to the stresses of a mission profile spanning several decades. In fact, this result should be placed in a realistic functional context. On the one hand, the nominal capacity of the cables [IEC 82] systematically exceeds the real load, and, on the other hand, the probability of a temporary overload is relatively low.

As an indication, for $\left(\overline{\theta}_c \approx 20°C \Leftrightarrow \rho \approx 6\right)$, the probability of failure would not exceed $\left(P_f \approx 0.3\%\right)$.

3.7. Conclusion and extensions

The generic character of the analytical model of *"exponential/normal"* interaction is justified through various arguments, which are physical, statistical and informational in nature.

In terms of design based on the *"stress–strength"* approach, it enables us to quickly determine the probability of failure as well as the safety factor of a device at the development stage.

In a context of poor information, when the mission profile that determines the constraints is uncertain, this offers the possibility of integrating "*arbitrary*" values by limiting their arbitrary character.

In conclusion, we note that this interaction model can potentially be used to optimize the reliability of a device, taking into account technical and economic constraints. Lind [LIN 93] has carried out a numerical integration study on the influence of maintenance costs and the discount rate on the probability of failure and the safety factor.

Taraman and Kapur [TAR 75] have proposed procedures for optimizing the parameters of the laws by considering policies for researching the minimum cost with an imposed reliability, or maximum reliability with an imposed cost.

3.8. References

[AIT 57] AITCHISON J., BROWN J.A.C., *The Lognormal Distribution*, Cambridge University Press, 1957.

[AME 61] AMERICAN SOCIETY FOR METALS, *Metals Handbook, 8th Edition, Volume 1 – Properties and Selection of Metals*, American Society for Metals, Metal Park, Ohio, 1961.

[AND 88] ANDERS G.J., FOTY M.S., CROALL S.C., "Probabilistic analysis of underground power cables", *Proc. of 2nd Int. Symp. on Applic. of Probability Methods in Power Systems*, San Francisco, California, 1988.

[BUR 78] BURY K.V., "On product reliability under random field loads", *IEEE Tr. on Reliab.*, vol. R-27, no. 4, pp. 258–260, 1978.

[DAV 58] DAVENPORT W.B., ROOT W.L., *An Introduction to the Theory of Random Signals and Noise*, McGraw-Hill, New York, 1958.

[FOX 87] FOX C., *An Introduction to the Calculus of Variations*, Dover Publication, Mineola, New York, 1987.

[GUM 58] GUMBEL E.J., *Statistics of Extremes*, Columbia University Press, New York, 1958.

[HAU 65] HAUGEN E.B., Statistical strength properties of common metals alloys, Space and Information Systems Division, North American Aviation Inc., SID 65-1274, 1965.

[IEC 82] IEC, Calculation of the Continuous Current Rating of Cables (100% load factor), IEC Publication 287, 2nd edition (also Cable Ampacity, IEC 60502, 2005), 1982.

[IND 84] INDULKAR C.S., "Sensitivity analysis of the ampacities of self-contained, oil-paper insulated, naturally cooled and forced cooled 300 kV underground cables", *Proc. IEEE Letter*, vol. 72, no. 2, pp. 233–234, 1984.

[JAY 57] JAYNES E.T., "Information theory and statistical mechanics", *Physical Review*, vol. 106, no. 4, pp. 620–630, May 15, 1957.

[LIN 93] LIND R.C., "Target reliability levels from social indicators", *Structural Safety and Reliability, Risk Assessment Session*, vol. 3, pp. 1897–1904, 1993.

[MIS 70] MISCHKE C., "A method relating factor of safety and reliability", *J. of Engrg. for Ind., Series B*, vol. 92, no. 3, pp. 537–542, 1970.

[PAP 84] PAPOULIS A., *Probability, Random Variables and Stochastics Processes*, McGraw-Hill, New York, 1984.

[PIE 92] PIERRAT L., "Estimation de la probabilité de défaillance par interaction de deux lois de Weibull", *Revue de statistique appliquée*, vol. XXXX, no. 4, pp. 5–13, 1992.

[PIE 93] PIERRAT L., "Le déséquilibre des réseaux et son influence sur les perturbations harmoniques", *Symposium CIGRE, Power System EMC, Session 1 (Sources of Disturbances)*, October 18–20, 1993.

[PIE 08] PIERRAT L., "Variabilité statistique de la vitesse du vent et potentiel énergétique d'un site éolien", *Revue essais industriels*, vol. 45, pp. 13–16, 2008.

[PIE 10] PIERRAT L., VANUXEEM J., Guidance for tailoring material to its life cycle environment profile (Mechanical Environment). Contributions to articles 5.6 (characterisation of variability) & 5.7 (interactions between Normal, Lognormal and Weibull distributions), Draft 0, ASTE, 2010.

[RAY 96] RAYLEIGH J.W.S., *The Theory of Sound*, vols 1–2 , MacMillan & Co., London, 1896.

[RIT 09] RITZ W., "Über eine neue Methode zur Lösung gewisser Variation probleme der mathematischen Physik", *J.f. reine und. angew. Math.*, vol. 135, no. 1, pp. 1–61, 1909.

[SCH 84] SCHUSTER E.F., "Classification of probability laws by tail behavior", *JASA, Theory and Methods Section*, vol. 79, no. 388, pp. 936–939, 1984.

[SHA 48] SHANNON C.E., "A mathematical theory of communication", *Bell System Technical Journal*, vol. 27, pp. 379–423/pp. 623–656, 1948.

[SPI 68] SPIEGEL M.R., *Mathematical Handbook of Formulas and Tables*, McGraw Hill, New York, 1968.

[TAR 75] TARAMAN S.I., KAPUR K.C., "Optimization considerations in design reliability by stress-strength interference theory", *IEEE Trans. on Reliab.*, vol. R-24, no. 2, pp. 136–138, 1975.

[WEI 51] WEIBULL W., "A statistical distribution function of wide applicability", *J. Appl. Mech., Am. Soc. Mech. Eng.*, vol. 18, pp. 293–297, 1951.

[WIR 76] WIRSCHING P.H., "On the behavior of statistical models used for design", *J. Eng. Ind.*, vol. 98, no. 2, pp. 601–606, available at: https://doi.org/10.1115/1.3438944, 1976.

[WIT 37] WITTAKER E.T., *A Treatise on the Analytical Dynamics of Particles and Rigid Bodies*, 4th edition, Cambridge University Press, Cambridge, 1937.

[ZIE 65] ZIENKIEWICZ O.C., CHEUNG Y.E., "Finite elements in the solution of field problems", *The Engineer*, pp. 507–510, 1965.

4

Reliable Optimization of Dental Implants Using the Generalized Polynomial Chaos Method

4.1. Introduction

Orthodontic prostheses can be used to replace or reconstruct teeth. The dental implant is usually composed of two parts: an artificial root that is fixed into the bone and comes up through the gum, and serves to support a prosthetic tooth (or element). Proos et al. [PRO 01] have studied the design of dental restorations. More recently, Couegnat et al. [COU 06] have studied shape optimization for cavity preparation. This study involves using shape optimization technology to design alternative shapes for dental implants so as to optimize stress distribution along the bone–implant interface.

In [HED 11], the optimization of the dental implant is performed using the concept of functionally graded materials. It is well known that a numerical study of such a dental implant, which does not take into account the uncertainties of various parameters, has attracted a lot of attention. Nevertheless, the system parameters may be impaired due to a lack of precision. In this context, the variability of the system parameters is a complex area that involves identifying and modeling the sources of uncertainty, the propagation of uncertainty and post-processing in order to measure the influence of this variability on the system behavior [DAM 17]. The main methods that include uncertainties are probabilistic with uncertain parameters that are modeled as random variables through probability density functions [BRU 07]. Among the probabilistic approaches, the Monte Carlo (MC) method [HAM 64] is

---

Chapter written by Fatma ABID, Abdelkhalak EL HAMI, Tarek MERZOUKI, Hassen TRABELSI, Lassaad WALHA and Mohamed HADDAR.

For a color version of all the figures in this chapter, see: http://www.iste.co.uk/elhami/applied1.zip.

widely used in this field. However, it does take a long time to obtain accurate results. Therefore, the MC method is used as a method to validate the effectiveness of other approaches such as in [GUE 16a] and [GUE 16b]. The spectral stochastic approach is also a solution for stochastic problems. Generalized polynomial chaos (GPC) is considered as one of the spectral stochastic methods. This approach is more efficient than other methods such as the MC method. In various works, the GPC method is used and compared with other uncertainty methods [SAH 13]. The originality of this paper is to increase the robustness of the dental implant that functions in the presence of uncertainties. Reliability-based design optimization (RBDO) is coupled with the stochastic approach in order to obtain the best structural design that guarantees the prerequisite robustness and level of reliability.

This chapter is organized into six sections, including the introduction and conclusion. In section 4.2, the description of the stochastic approach, such as the MC method and the GPC method, is presented. RBDO methods are presented in section 4.3. Finally, the numerical results are detailed in section 4.4.

4.2. Stochastic approach

Different techniques exist to model the propagation of uncertainty. These methods are classified into three categories: simulation methods, perturbation methods and spectral methods. The MC method is considered as a reference method for the calculation of systems with uncertain parameters. The major disadvantage of this method is the very large amount of computation that complicates its use.

The perturbation method is based on a Taylor series development for the response around its mean. The main problem of this method comes from the conditions, which ensure the convergence of these series. The variables must have a weak distribution. The method adopted to take into account the uncertainty is the GPC method. The fundamental idea of the GPC method, invented by Norbert Wiener in 1938 [WIE 38], is to establish a separation between the stochastic components of the random function and its deterministic components.

4.2.1. *The MC method*

In this section, the theory of the MC method is presented. Spanos and Kontsos [SPA 08] have delivered an exposition on this approach. This method provides successive resolutions of a deterministic system that incorporates random variables. This method is used when the problem to be treated is complex. It generates random samples for all uncertain parameters and according to their probability distributions [GUR 15]. For each sample, a set of parameters is obtained and a deterministic calculation followed by a numerical or analytical model is made. The main advantage of this technique is its ease of implementation. Whatever the size and complexity of a system, the results of such a method are accurate. However, it is often expensive, as any reasonable accuracy requires a large number of samples.

In this work, the MC method is considered as a reference method to validate the effectiveness of GPC.

The functions of the MC method are as follows:

$$Y = M(X) \quad [4.1]$$

where X represents the vector for uncertain input parameters modeled by a random vector $X = [X_1; X2; \ldots; X_n]^T$, where M is the model considered, and Y represents the vector of estimated outputs which will be a random vector.

4.2.2. *The GPC method*

This section describes the GPC method. A demonstration of this method is given by Sepahvand et al. [SEP 15]. GPC provides, for any stochastic process, a linear combination of the orthogonal polynomials of random variables. The fundamental idea of the GPC method was developed by Wiener and elaborated by [BRU 07] Bruyère, who used GPC to deal with general random fields, such as multidimensional random fields and non-Gaussian random fields.

Consider a set of random variables (X_i) that can be written as a second-order random process. Therefore, ($X_{i,}$) can be expressed in the following form:

$$X_{i,j}(x,\xi) = \sum_{j=0}^{\infty} Xb_{i,j}(x)\ \psi_j(\xi) \quad [4.2]$$

where ξ is the vector of random variables, $Xb_{i,j}$ is the deterministic component and ψ_j (ξ) are the orthogonal polynomial functions such as Legendre, Hermite, etc.

Orthogonal polynomials verify the orthogonality relation:

$$<\psi_i, \psi_j> = \int \psi_i\ \psi_j\ W(\xi)\ d\xi = \begin{cases} 0\ \mathit{if}\ \ i \neq j \\ <\psi_i, \psi_j>\ \mathit{if}\ \ i = j \end{cases} \quad [4.3]$$

W (ξ) is the probabilistic density function and <,> is the inner product operator. In practice, the GPC expansion is truncated to a finite order P in order to limit the number of terms in the sum to a finite number. This truncation is essential for the problem to remain feasible.

$$X_{i,j}(x,\xi)=\sum_{j=0}^{p} Xb_{i,j}(x)\ \psi_j(\xi) \qquad [4.4]$$

with

$$P=\frac{(p+r)!}{p!r!}-1 \qquad [4.5]$$

Such that p is the order of the polynomial chaos and r is the dimension of the random vector in the system. The calculation of the representation by the GPC method requires the determination of $P + 1$ deterministic components called stochastic modes. Two approaches are needed to determine such modes: intrusive and non-intrusive. The first approach consists of substituting random functions, in the stochastic model, by their spectral representation. Next, a projection called Galerkin of the resulting system on GPC is made [DAM 17]. The second approach consists of directly projecting the stochastic solution from the GPC. A non-intrusive approach may be more attractive for a stochastic system since it requires little computation for deterministic solutions at a finite number of points in the domain of probabilistic variation.

4.3. Deterministic design optimization

The use of deterministic design optimization (DDO) has become common in structural design with the development of finite element (FE) software. This method consists of adding safety factors to the constraint function so as to incorporate uncertainties into the design variables [YAI 17]. Therefore, the optimization problem consists of minimizing the objective function $f(x)$ subjected to different constraints such as geometric, physical or functional constraints.

$$\begin{aligned}&\min_{x} : f(x)\\ &t.q: g_k(x,\xi)\le 0 \quad k=1,\dots,K\end{aligned} \qquad [4.6]$$

where x is the vector of deterministic design variables, ξ is the vector of random variables and $g_k(x, \xi)$ are the inequality constraints. In reality, designers cannot select the appropriate safety factors to increase safety as much as possible.

To overcome the DDO problem, RBDO is essential to ensure an optimal and reliable structural design.

4.4. Reliability-based design optimization

The RBDO procedure seeks the best compromise between cost and safety, taking into account the uncertainties related to the system. The basic formulation of RBDO consists of minimizing the objective function under probabilistic constraints. In RBDO, two types of variables are taken into account: design variables x and random variables y. The typical RBDO formulation is written as [ELH 13]:

$$\min_{x} : f(x)$$

$$t.q: \begin{cases} \Pr\left[G_i(x,y) \leq 0\right] \leq P_{f_i}^T & i = 1,\ldots,r \\ h_k(x) \leq 0 & k = r+1,\ldots,R \end{cases} \qquad [4.7]$$

where x is the vector of design variables, y is the vector of random variables, $f(x)$ is the objective function, $G_i(x,y)$ is the performance function, $h_k(x)$ are deterministic constraints, Pr [.] is the probability operator, $P^T{}_{fi}$ is the probability of acceptable failure, r is the number of runtime functions and R is the total number of constraints.

Note that the design variables x can be deterministic variables, system parameters (e.g. materials, loads and dimensions) or probabilistic distribution parameters (e.g. mean values and standard deviations of random variables y) [MAK 16]. The random variables y can be geometric dimensions or applied loads. The uncertainties of these variables can be modeled by statistical information. The deterministic constraints h_k are the upper and lower bounds of the design variables.

In the RBDO approach, the reliability index is calculated iteratively. Therefore, solving an RBDO problem requires a lot of computational efforts in order to solve such a design. In order to reduce the computational time, the hybrid method (HM) was developed. It aims at solving the optimization and reliability evaluation simultaneously. By doing this, the optimization problem becomes more complex than the deterministic problem. An efficient method called optimal safety factor (OSF) has been developed. By using this method, the computation time is reduced when compared to the HM.

4.4.1. *The classic method*

In the classical approach, we use two spaces, namely the physical space and the normalized space [ELH 13]. Therefore, classical reliability optimization is performed by considering two nested problems:

– The optimization problem:

This aims to minimize the objective function $f(x)$ subject to both the deterministic constraints $g_l(x)$ and the reliability constraint. It can be defined as follows:

$$\min_{x}: \ f(x) \qquad t.q: \begin{cases} g_l(x) \leq 0 \quad l = 0..L \\ \beta(x,u) \geq \beta_t \end{cases} \tag{4.8}$$

where $f(x)$ is the objective function, $g_l(x)$ is the associated constraint, $\beta(x, u)$ is the reliability index of the structure and β_t is the reliability target.

– The reliability problem:

The reliability index $\beta(x, u)$ is the minimum distance between the limit state function $H(x, u)$ and the origin. This index is determined by solving the minimization problem:

$$\min_{u}: \ d_\beta(u) \qquad t.q: \ H(x,u) \leq 0 \tag{4.9}$$

where $d_\beta(u)$ is the distance in the normalized random space, given by $d_\beta(u) = \sqrt{\Sigma u_i^2}$, and $H(x,u)$ is the performance function (or limit state function) in the normalized space.

It has been shown that the classical approach requires a high computational time due to the large number of computations in both spaces and can lead to a low stability in terms of the convergence. To overcome this problem, a new method called OSF has been developed [KHR 08].

4.4.2. *OSF using GPC*

In the case of DDO, designers choose safety factors based on their engineering experience. Therefore, the safety of the design is not guaranteed and the optimal point may not lead to the required level of reliability. Nevertheless, in the case of the OSF approach, the determination of safety factors is independent of engineering experience. As such, the OSF approach is the best method to deal with this problem and also to reduce the computational time [ELH 13].

The OSF algorithm using the GPC method can be implemented in three steps. The proposed approach is described in the flowchart shown in Figure 4.1.

1) Determine the point of design:

This step consists of solving the optimization problem by minimizing the objective function that is subject to the limit state function, deterministic and probabilistic constraints, computed using the GPC. The resulting solution is designed by the design point, also called the most probable failure point, as well as called the design point y^*_i.

2) Calculate the safety factors:

The design point y^*_i can be related to the optimal value x^*_i by equation 4.13.

$$y_i^* = S_f^i \, x_i^*, \quad i=1,...,n \qquad [4.10]$$

Moreover, in general, when considering the normal distribution law of the random variable y_i, the transformation from physical space (x-space) to normalized space (u-space) is given by:

$$u_i = \frac{y_i - x_i}{\sigma_i}, \quad i=1,...,n \qquad [4.11]$$

where σi is the standard deviation. In this way, we define the safety factors S^i_f, which can be expressed as:

$$S_f^i = 1 + \gamma_i \, u_i^* \quad i=1,..,n \qquad [4.12]$$

where γ_i is the coefficient relating the standard deviation and the mean value for the random variable m_i. Therefore, γ_i can be expressed as follows:

$$\gamma_i = \frac{\sigma_i}{m_i} \qquad [4.13]$$

The equation for the optimal value of the normalized vector u^*_i can be written in the following form:

$$u_i^* = \pm\beta_t \sqrt{\frac{\left|\frac{\partial G}{\partial y_i}\right|}{\sum_{i=1}^{n}\left|\frac{\partial G}{\partial y_i}\right|}} \quad i=1,\ldots,n \qquad [4.14]$$

The ± sign depends on the sign of the derivative:

$$\frac{\partial G}{\partial y_i} > 0 \rightarrow u_i^* > 0 \ et \ \frac{\partial G}{\partial y_i} < 0 \rightarrow u_i^* < 0 \quad i=1,..,n \qquad [4.15]$$

3) Calculate the optimal solution:

In this last step, we determine the optimal design by including the calculated safety factors in the values of the design variables.

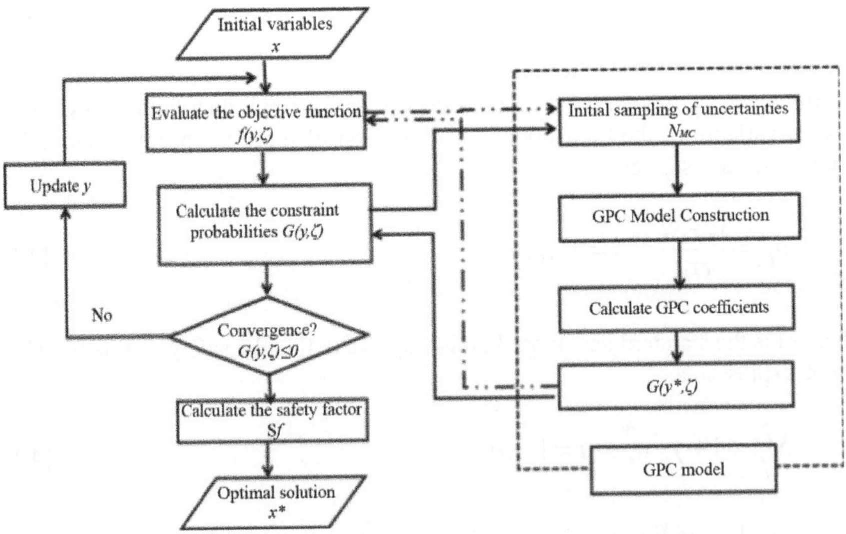

Figure 4.1. *Flowchart of the OSF function using the GPC metamodel*

The optimization problem is solved by the sequential quadratic programming (SQP) tool using MATLAB software.

4.5. Numerical result

In this section, a numerical example of a 2D dental implant is studied. A probabilistic analysis is implemented in order to study the robustness of the system. Next, a reliable optimization using GPC is performed to obtain the optimal design of the structure.

4.5.1. *2D dental implant*

Figure 4.2(a) shows a simplified geometric model of the dental implant under study. We consider two variables r and a, which, respectively, represent the radius of curvature at the pointed part of the implant and the radius of the implant. The material of the implant is a titanium alloy. The dimensions and properties are outlined in Table 4.1. The numerical simulation of the dental implant is performed using ANSYS software. Regarding the discretization of the structure, there are 2,323 elements assigned to the structure.

The adopted boundary conditions are, respectively, the forces F_x and F_y applied to the upper part of the dental implant as shown in Figure 4.2(b). All the nodes at the edge of the structure are constrained.

a) b)

Figure 4.2. *(a) Geometric model of the dental implant; (b) boundary conditions applied to the dental implant*

Description	Parameters	Value
Radius of curvature	r	2.1 mm
Radius of the implant	a	1.83 mm
Young's modulus	E	193,000 MPa
Poisson's ratio	v	0.31
Axial force	Fx	30 N
Transversal force	Fy	30 N

Table 4.1. *Dimensions and properties of the system*

At the initial point, the values of the optimization variables are: $r = 2.1$ mm and $a = 1.83$ mm, which leads to an initial value of the structural volume equal to $V_{initial} = 55.68$ mm^3. The maximum value of the Von–Mises stress is equal to $\sigma_{max} = 177.18$ MPa.

For the optimization problem, the objective is to minimize the volume V of the dental implant that is subjected to the probabilistic stress in relation to the permissible stress condition. We consider that the design points are associated with the radius of curvature and the radius of the implant.

To perform the deterministic optimization, a safety factor $S_f = 1.25$ is proposed. The mathematical model of the DDO is given by:

$$\begin{aligned} &\min_{x} : V(x,\xi) \\ &t.q : \sigma_{\max}(x,\xi) \leq \frac{\sigma_w}{S_f} \end{aligned} \quad [4.16]$$

where σ_{max} is the maximum value of the Von–Mises stresses and σ_w is the allowable stress equal to 185 MPa. To perform the reliability optimization using the OSF method, the first step is to obtain the design point. The formulation of the optimization problem is given as:

$$\begin{aligned} &\min_{x} : \mathrm{V}(x,\xi) \\ &t.q : \sigma_{\max}(x,\xi) \leq \sigma_w = 185\mathrm{MPa} \end{aligned} \quad [4.17]$$

The next step is to calculate the safety factors and finally determine the optimal solution by dividing the design point value by the calculated safety factor.

4.5.1.1. *Probabilistic analysis*

In this section, the discussion on the variability of uncertain parameters in the dental implant is introduced. The objective of this section is to improve the robustness and performance of the dental implant operating in the presence of uncertain parameters of the material properties.

These parameters are chosen at random and follow a uniform distribution around their nominal values: 1%. Table 4.4 presents an idea of the various uncertain parameters selected for this example of the dental implant technique.

Parameters	**E (MPa)**	**ν**
Value	U [191070,194930]	U [0.3069,0.3131]

Table 4.2. *Characteristics of uncertain parameters*

Processing the MC method consists of creating a grid of numerical values from the uncertain parameters and then calculating the quantity of interest for the system. The quantity of interest is analyzed for 1,000 simulations. Figure 4.3 shows the distribution of the input variables (E, v) in the case of a uniform distribution.

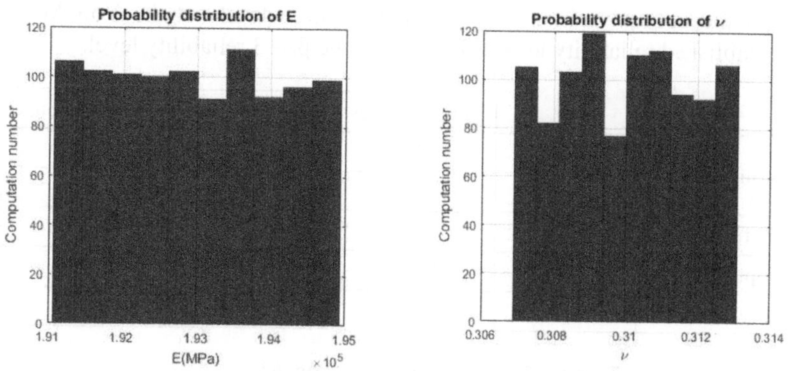

a) b)

Figure 4.3. *(a) Geometric model of the dental implant; (b) boundary conditions applied to the dental implant*

4.5.1.2. *RBDO using GPC*

To obtain the best design of the structure, we introduce the RBDO approach associated with the FE simulation of the dental implant.

4.5.1.2.1. DDO approach using the GPC model

Figure 4.5 shows the evolution of the objective function with the iterations.

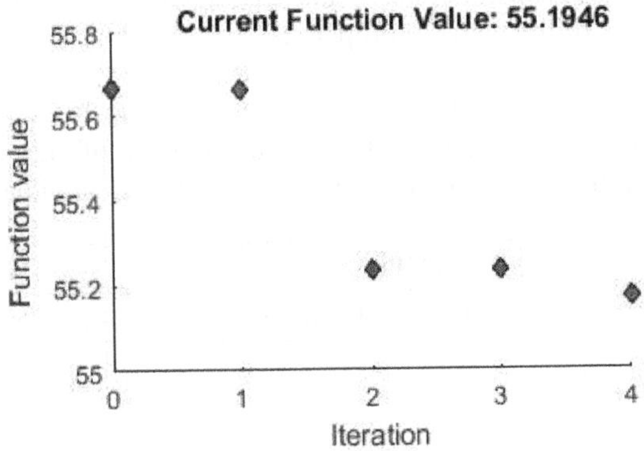

Figure 4.4. *Iterative curve of the dental implant volume by the DDO method*

The results of this technique are shown in Table 4.3. It is necessary to mention that although the optimal solution satisfies the required levels of the Von–Mises stress, the established reliability level is lower than the $\beta_t = 3$ reliability level.

		Initial values	Design point	Optimal point
Design variables	r (mm)	2.1	2.13	2.12
	a (mm)	1.83	1.8	1.8
Objective function	V (mm^3)	55.68	55.17	55.19
Constraint optimization	σ_{max} (MPa)	179.54	175.60	177.18
Reliability index	β	–	–	1.18

Table 4.3. *DDO using GPC results*

For this reason, the following method is used to ensure the robustness and reliability of our system.

4.5.1.2.2. DDO approach using the GPC model

Figure 4.5 illustrates the evolution of the objective function with iterations, while Table 4.4 shows the results of this approach.

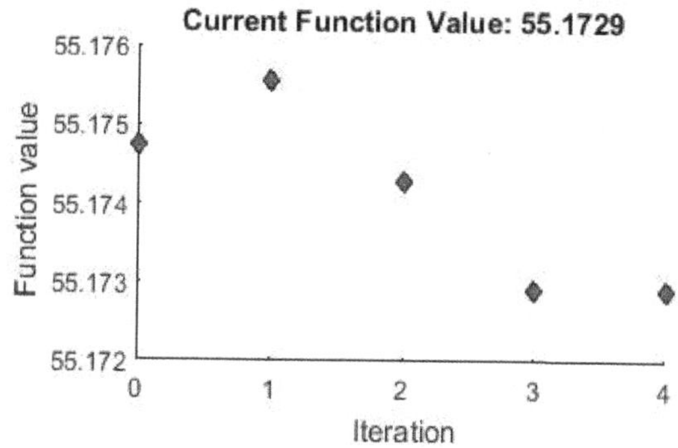

Figure 4.5. *Iterative curve of the dental implant volume through the OSF method*

		Initial values	Design point	OSF Sf	Optimal point
Design variables	r (mm)	2.13	2.13	0.84	2.52
	a (mm)	1.8	1.8	1.04	1.72
Objective function	V (mm^3)	55.17	55.17		53.75
Constraint optimization	σ_{max} (MPa)	175.60	175.60		169.91
Reliability index	β	–	–		3

Table 4.4. *OSF using GPC results*

It should be noted that with the OSF approach, the computational cost is low when compared to other RBDO methods, such as the HM approach ([ELH 13a] and [ELH 13b]). Table 4.5 shows a comparison of the optimal point, between the DDO method and the OSF method using the GPC metamodel.

	DDO	OSF
r (mm)	2.12	2.52
a (mm)	1.8	1.72
Volume (V)	55.19	53.75
Constraint optimization	177.18	169.91
Reliability index	1.18	3

Table 4.5. *Comparison of the optimal point between DDO and OSF*

From Table 4.5, we can see that:

– for the deterministic optimization using the GPC model, the maximum tension stress is 121.39, which satisfies the stress condition (less than $\sigma_{max} = S_f$) and the structural volume is 55.19, but the reliability level is low ($\beta = 1.18$), which does not guarantee the required reliability target;

– for the OSF using the GPC model, the maximum Von–Mises stress is 169.91, which satisfies the stress condition (less than 185 MPa) and the structural volume is 53.75.

4.6. Conclusion

In this chapter, we apply the GPC method to dental implant solutions when the material properties have uncertainties. In our study, we propose to take into account the propagation of uncertainty by combining the GPC-based metamodel with the OSF method.

The numerical example demonstrated the effectiveness and efficiency of an OSF method based on the GPC model for the reliability optimization of a dental implant with random variables.

4.7. References

[ABO 11a] ABO AL-KHEER A., EID M., AOUES Y., et al., "Theoretical analysis of the spatial variability in tillage forces for fatigue analysis of tillage machines", *Journal of Terramechanics*, vol. 48, no. 4, pp. 285–295, 2011.

[ABO 11b] ABO AL-KHEER A., KHARMANDA M.G., EL HAMI A., et al., "Estimating the variability of tillage forces on a chisel plough shank by modeling the variability of tillage system parameters", *Computers and Electronics in Agriculture*, vol. 78, no. 1, pp. 61–70, 2011.

[ABO 11c] ABO AL-KHEER A., EL HAMI A., KHARMANDA M.G., et al., "Reliability-based design for soil tillage machines", *Journal of Terramechanics*, vol. 48, no. 1, pp. 57–64, 2011.

[BEN 13] BEN ABDESSALEM A., PAGNACCO E., EL HAMI A., "Increasing the stability of T-shape tube hydroforming process under stochastic framework", *International Journal of Advanced Manufacturing Technology*, vol. 69, nos 5–8, pp. 1343–1357, 2013.

[BEN 15] BEN ABDESSALEM A., EL HAMI A., "A probabilistic approach for optimising hydroformed structures using local surrogate models to control failures", *International Journal of Mechanical Sciences*, vols 96–97, pp. 143–162, 2015.

[BOU 96] BOUYEKHF R., EL MOUDNI A., EL HAMI A., et al., "Reduced order modelling of two-time-scale discrete non-linear systems", *Journal of the Franklin Institute*, vol. 333, no. 4, pp. 499–512, 1996.

[BOU 01] BOUYEKHF R., EL HAMI A., EL MOUDNI A., "Optimal control of a particular class of singularly perturbed nonlinear discrete-time systems", *IEEE Transactions on Automatic Control*, vol. 46, no. 7, pp. 1097–1101, 2001.

[BRU 07] BRUYÈRE J., DANTAN J.-Y., BIGOT R., et al., "Statistical tolerance analysis of bevel gear by tooth contact analysis and Monte Carlo simulation", *Mechanism and Machine Theory*, vol. 42, no. 10, pp. 1326–1351, 2007.

[CHR 08] CHEROUAT A., RADI B., EL HAMI A., "The frictional contact of the composite fabric's shaping", *Acta Mechanica*, vol. 199, nos 1–4, pp. 29–41, 2008.

[COU 06] COUEGNAT G., FOK S.L., COOPER J.E., et al., "Structural optimization of dental restorations using the principle of adaptive growth", *Dental Materials*, vol. 22, no. 1, pp. 3–12, 2006.

[DAM 17] DAMMAK K., EL HAMI A., KOUBAA S., et al., "Reliability based design optimization of coupled acoustic-structure system using generalized polynomial chaos", *International Journal of Mechanical Sciences*, vol. 134, pp. 75–84, 2017.

[DAM 19] DAMMA K., KOUBAA S., EL HAMI A., et al., "Numerical modelling of vibro-acoustic problem in presence of uncertainty: Application to a vehicle cabin", *Applied Acoustics*, vol. 144, pp. 113–123, 2019.

[DEB 21] DEBICH B., Modélisation, simulation, et fiabilité des matériaux changement de phase pour l'automobile et l'aéronautique, Thesis, INSA Rouen Normandie, 2021.

[ELH 93] EL HAMI A., LALLEMENT G., MINOTTI P., et al., "Methods that combine finite group theory with component mode synthesis in the analysis of repetitive structures", *Computers and Structures*, vol. 48, no. 6, pp. 975–982, 1993.

[ELH 96] EL HAMI A., RADI B., "Some decomposition methods in the analysis of repetitive structures", *Computers and Structures*, vol. 58, no. 5, pp. 973–980, 1996.

[ELH 09] EL HAMI A., RADI B., CHEROUAT A., "Treatment of the composite fabric's shaping using a Lagrangian formulation", *Mathematical and Computer Modelling*, vol. 49, nos 7–8, pp. 1337–1349, 2009.

[ELH 11] EL HAMI A., RADI B., "Comparison study of different reliability-based design optimization approaches", *Advanced Materials Research*, vol. 274, pp. 113–121, 2011.

[ELH 13a] EL HAMI A., RADI B., *Incertitude et optimisation et fiabilité des structures*, Hermes-Lavoisier, Paris, 2013.

[ELH 13b] EL HAMI A., RADI B., *Uncertainty and Optimization in Structural Mechanics*, ISTE Ltd, London and John Wiley & Sons, New York, 2013.

[GUE 15a] GUERINE A., EL HAMI A., WALHA L., et al., "Approach for the dynamic analysis of one stage gear system with uncertain parameters", *Mechanism and Machine Theory*, vol. 92, pp. 113–126, 2015.

[GUE 15b] GUERINE A., EL HAMI A., FAKHFAKH T., et al., "A polynomial chaos method to the analysis of the dynamic behavior of spur gear system", *Structural Engineering and Mechanics*, vol. 53, no. 4, pp. 819–831, 2015.

[GUE 16a] GUERINE A., EL HAMI A., WALHA L., et al., "A polynomial chaos method for the analysis of the dynamic behavior of uncertain gear friction system", *European Journal of Mechanics, A/Solids*, vol. 59, pp. 76–84, 2016.

[GUE 16b] GUERINE A., HAMI A.E., WALHA L., et al., "Dynamic response of a Spur gear system with uncertain friction coefficient", *Advances in Engineering Software*, vol. 120, pp. 45–54, 2016.

[GUE 17] GUERINE A., EL HAMI A., WALHA L., et al., "Dynamic response of wind turbine gear system with uncertain-but-bounded parameters using interval analysis method", *Renewable Energy*, vol. 113, pp. 679–687, 2017.

[HAM 64] HAMMERSLEY J., HANDSCOMB D., *Monte Carlo Methods: Methuen's Monographs on Applied Probability and Statistics*, Methuen, London, 1964.

[HED 11] HEDIA H.S., "Numerical optimization of dental implant through the concept of functionally graded materials", *Materials Testing*, vol. 53, no. 9, pp. 556–563, 2011.

[JAN 17] JANNOUN M., AOUES Y., PAGNACCO E., et al., "Probabilistic fatigue damage estimation of embedded electronic solder joints under random vibration", *Microelectronics Reliability*, vol. 78, pp. 249–257, 2017.

[KHA 08] KHARMANDA G., MOHSINE A., MAKLOUFI A., et al., "Recent methodologies for reliability-based design optimization", *International Journal for Simulation and Multidisciplinary Design Optimization*, vol. 2, no. 1, pp. 11–23, 2008.

[KHA 09] KHARMANDA G., SHARABATEY S., IBRAHIM H., et al., "Reliability-based design optimization using semi-numerical strategies for structural engineering applications", *International Journal of CAD/CAM*, vol. 9, no. 1, pp. 1–16, 2009.

[KUS 14] KUSANO I., BALDOMIR A., JURADO J.A., et al., "Reliability based design optimization of long-span bridges considering flutter", *Journal of Wind Engineering and Industrial Aerodynamics*, vol. 135, pp. 149–162, 2014.

[LED 17a] LEDOUX M., EL HAMI A., *Fluid Mechanics: Analytical Methods*, ISTE Ltd, London and John Wiley & Sons, New York, 2017.

[LED 17b] LEDOUX M., EL HAMI A., *Compressible Flow, Propulsion and Digital Approaches in Fluid Dynamics*, ISTE Ltd, London and John Wiley & Sons, New York, 2017.

[LED 21a] LEDOUX M., EL HAMI A., *Heat Transfer 1. Conduction*, ISTE Ltd, London and John Wiley & Sons, New York, 2021.

[LED 21b] LEDOUX M., EL HAMI A., *Heat Transfer 2. Radiative Transfer*, ISTE Ltd, London and John Wiley & Sons, New York, 2021.

[MAB 17] MABROUK I.B., EL HAMI A., WALHA L., et al., "Dynamic vibrations in wind energy systems: Application to vertical axis wind turbine", *Mechanical Systems and Signal Processing*, vol. 85, pp. 396–414, 2017.

[MAK 16] MAKHLOUFI A., AOUES Y., EL HAMI A., "Reliability based design optimization of wire bonding in power microelectronic devices", *Microsystem Technologies*, vol. 22, no. 12, pp. 2737–2748, 2016.

[MOH 10] MOHSINE A., EL HAMI A., "A robust study of reliability-based optimization methods under eigen-frequency", *Computer Methods in Applied Mechanics and Engineering*, vol. 199, nos 17–20, pp. 1006–1018, 2010.

[MOR 02] MORO T., EL HAMI A., EL MOUDNI A., "Reliability analysis of a mechanical contact between deformable solids", *Probabilistic Engineering Mechanics*, vol. 17, no. 3, pp. 227–232, 2002.

[PRO 01] PROOS K., STEVEN G., SWAIN M., et al., "Preliminary studies on the optimum shape of dental bridges", *Computer Methods in Biomechanics and Biomedical Engineering*, vol. 4, no. 1, pp. 77–92, 2001.

[RAD 07] RADI B., EL HAMI A., "Reliability analysis of the metal forming process", *Mathematical and Computer Modelling*, vol. 45, nos 3–4, pp. 431–439, 2007.

[SAH 13] SAHA S., SEPAHVAND K., MATSAGAR V., et al., "Stochastic analysis of base-isolated liquid storage tanks with uncertain isolator parameters under random excitation", *Engineering Structures*, vol. 57, pp. 465–474, 2013.

[SAR 11] SARSRI D., AZRAR L., JEBBOURI A., et al., "Component mode synthesis and polynomial chaos expansions for stochastic frequency functions of large linear FE models", *Computers and Structures*, vol. 89, nos 3–4, pp. 346–356, 2011.

[SEP 15] SEPAHVAND K., SCHEFFLER M., MARBURG S., "Uncertainty quantification in natural frequencies and radiated acoustic power of composite plates: Analytical and experimental investigation", *Applied Acoustics*, vol. 87, pp. 23–29, 2015.

[SPA 08] SPANOS P., KONTSOS A., "A multiscale Monte Carlo finite element method for determining mechanical properties of polymer nanocomposites", *Probabilistic Engineering Mechanics*, vol. 23, no. 4, pp. 456–470, 2008.

[WIE 38] WIENER N., "The homogeneous chaos", *American Journal of Mathematics*, vol. 60, no. 4, pp. 897–936, 1938.

[XIA 15] XIA B., YU D., "Optimization based on reliability and confidence interval design for the structural-acoustic system with interval probabilistic variables", *Journal of Sound and Vibration*, vol. 336, pp. 1–15, 2015.

5

Multi-objective Reliability Optimization Based on Substitution Models. Applied Case Study of a Hip Prosthesis

5.1. Introduction

Traditional deterministic design optimization (DDO) does not take into account the presence of uncertainties in the system. As a result, the optimal point could easily violate the imposed stresses, resulting in an unreliable solution. In order to incorporate these uncertainties into the problem, reliability-based design optimization (RBDO) has been introduced by attracting more and more attention from researchers. Compared to DDO, RBDO aims at obtaining a reliable optimum by converting deterministic constraints into probabilistic constraints, in which the probability of failure is limited to a target level. In this context, several methods have been developed to solve this RBDO problem. In order to reduce the computational cost of the classical RBDO approach, [KHA 08] developed an efficient approach called the Hybrid Method (HM). In [MOH 10], the authors proposed an extension of the HM, called the improved HM, in order to improve the optimal value of the objective function.

This technique has significantly reduced the computational time; however, the optimization problem becomes more complex due to a large number of design variables. In [DAM 17], the authors incorporated the optimal safety factor

---

Chapter written by Khalil DAMMAK and Abdelkhalak EL HAMI.

For a color version of all the figures in this chapter, see: http://www.iste.co.uk/elhami/applied1.zip

associated with Generalized Polynomial Chaos in the study of a coupled acoustic-structure system. [YAI 18] and [DAM 17] presented a new extension of HM, called the robust HM (RHM) applied to an example structure subjected, respectively, to multiaxial random vibrations, and to a coupled acoustic-structure system. [YAI 17] developed the RHM for cases where structures are subjected to random vibrations. In [DAM 18], the authors applied the RHM to a coupled acoustic-structure system and validated the effectiveness of the method with different initial points. Debich et al. [DEB 22] investigated the RHM for the case of Phase-Change Materials (PCM)-based heatsink for cooling electronic boards. [KHA 09] developed a hybrid modified method for the case of an offshore wind turbine to overcome the shortcomings of other methods. Moreover, real engineering problems are usually characterized by more than one conflicting objective. For this reason, an appropriate trade-off between objective functions must be made using multi-objective optimization (MOO) [CHO 12, LIA 08, BOU 09]. Many authors have studied the MOO of structural-acoustic problems (e.g. [DAM 19, WAN 17]). Classical approaches aim at converting these objectives into a single-objective optimization problem by combining all objective functions into one. The difficulty with these classical approaches is that they require engineering expertise and the adopted algorithm must be applied repeatedly to obtain Pareto optimal solutions. Over the past decade, evolutionary algorithms (EA) have come to the forefront for solving MOO, due to their ability to obtain multiple Pareto optimal solutions in a single run [KAM 22]. Evolutionary multi-objective (EMO) optimization has become a popular research area whose usefulness has been validated through several industrial applications. Indeed, EA are based on the population concept in which several solutions participate in an iteration and develop a new population of solutions at each iteration. Due to this treatment of populations, EMO can simultaneously find several non-dominated solutions, which thus represent a trade-off between objectives, in a single iteration.

To solve these multi-objective engineering problems in the face of uncertainties, Barakat et al. [BAR 04] presented a general approach for performing reliability-based multi-objective design optimization (RBMDO) on prestressed concrete beams. In [DAS 11], the authors examined the effects of uncertainty on MOOs when applied to the design of aircraft wings. Fang et al. [FAN 15] proposed a MOO to improve both the performance and the robustness of the fatigue life of a truck. In [DEB 07], the authors combined traditional reliability optimization approaches and EMO optimization to better manage the uncertainties through the example of vehicle design.

Typically, when using computationally expensive simulation codes in complex engineering problems, it becomes impractical to perform a large number of simulations for uncertainty quantification or design optimization purposes. A better alternative is to use approximations of the original models, often called metamodels (or surrogate models). These metamodels aim to build a mathematical model in order to define the relationship between specific inputs and outputs of the system [FOR 09]. Surrogate models have been developed primarily to approximate deterministic simulations. Recent developments have explored their use in probabilistic analysis [DEY 17, ABI 20] and design optimization [DEB 11], LAU 13, DAM 21]. The most popular metamodeling methods are response surface methodology (RSM); for further reading, refer to some previous studies [KUR 02, GOE 07, DEY 15, MYE 16, SIM 01, MUK 17, ZHA 18].

5.2. Description of metamodeling methods

5.2.1. *Application of a substitution model*

The implementation of a substitution model is a process based on several steps. It uses several techniques such as those presented by Forrester et al. [FOR 08]. The schematic representation of the process for developing a substitution model is presented in Figure 5.1 and consists of five steps:

– a sampling procedure to obtain a set of points in the design space by the Design of Experiments (DOE) method;

– an evaluation procedure for the exact responses of the considered function to the points obtained, either through the Sampling Strategy or by the Enrichment procedure; this phase calls upon the considered mechanical model and finalizes the construction of the learning base E_a composed of a couple of responses (x_i, y_i);

– the metamodel construction procedure itself. This requires the use of methodologies specific to the substitution model being employed; this construction phase can be particularly long for some substitution models;

– validation of the metamodel by measuring its accuracy based on the minimization of the error between the substitution model and the exact model by a few points;

– an enrichment procedure (if convergence is not assured) allowing to add new sites to the learning base.

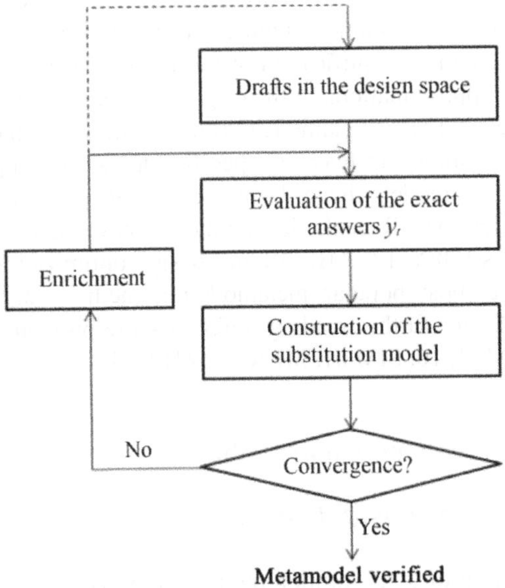

Figure 5.1. *Flowchart showing the developmental process of the substitution model*

5.2.2. Construction of a metamodel

Metamodel techniques are widely used today in engineering as technical analysis involves the implementation of a complex computer code. Metamodels are used to build approximations of the analysis code and describe the functional relationship between design variables and their corresponding responses. In this way, the computational cost is reduced.

The number of existing metamodels in the literature is quite large. However, they can be classified into two distinct categories:

– Parametric metamodels, as their name indicates, require the choice of internal parameters, either manually by a more or less empirical choice of the user, or automatically by implementing a methodology to fix them. For some parametric metamodels, the precise choice of these parameters is essential and cannot be ignored. These techniques, often specific to the type of metamodels employed, require adapted tools such as, for example, an optimization algorithm. This phase of choosing the parameter(s) can take up most of the cost needed to build the substitution model. Parametric models include Kriging, radial basis functions (RBF) and support vector regressions.

– Non-parametric metamodels that, unlike the previous ones, do not require any determination of internal parameters. Historically, this type of model was the first to be used, for example linear, quadratic and polynomial regression.

The two types of substitution models, polynomial response surface (PRS) and the Kriging method, used in this chapter, are described in the following sections.

To begin with, consider a dimensional problem *m*. We use $\boldsymbol{x} = \{x_1, x_2, ..., x_m\}$ to denote the design variables with dimensions m, $\boldsymbol{x_S} = \{x^{(1)}, x^{(2)}, ..., x^{(n)}\}^T$ to denote a sample set with n design points and $\boldsymbol{y_S} = \{y^{(1)}, y^{(2)}, ..., y^{(n)}\}^T = \{y(x^{(1)}), y(x^{(2)}), ..., y(x^{(n)})\}^T$ to show the corresponding output responses of the sample set. The pair $(\boldsymbol{x_S}, \boldsymbol{y_S})$ constitutes the sampled data set. The objective of metamodeling is to construct a surrogate model to predict the output of any untried point $\boldsymbol{x}$, in order to estimate $y(\boldsymbol{x})$ based on the pair $(\boldsymbol{x_S}, \boldsymbol{y_S})$.

The surrogate modeling approach first applies a DOE, such as Latin hypercube sampling (LHS) [MCK 00], to repeat the sampling in the design space, and then chooses a type of metamodel to represent the sampled data. Therefore, the original analysis code can be replaced by the surrogate model during the optimization process.

5.2.2.1. *Response surface methodology*

In a statistical and mathematical context, RSM is described by Myers et al. [MYE 16] as a set of methods for developing, improving, and optimizing processes and products. Using the Taylor series development, the PRS approximation has a polynomial form. In a PRS, the number of terms depends on the desired accuracy and the shape of the real function. It is defined as the approximation of the model $y = \hat{y}(\boldsymbol{x}) + \varepsilon$, where $\hat{y}(\boldsymbol{x})$ is the predictor, and where ε indicates the normal random error with zero mean and a variance of σ^2. The most used PRS is the second order or quadratic response surface (QRS) whose approximation $\hat{y}(\boldsymbol{x})$ can be expressed as follows:

$$\hat{y}(\boldsymbol{x}) = \beta_0 + \sum_{i=1}^{m} \beta_i x_i + \sum_{i=1}^{m} \sum_{j>i}^{m} \beta_{ij} x_i x_j + \sum_{i=1}^{m} \beta_{ii} x_i^2 \qquad [5.1]$$

Where β_0, β_i, β_{ij} shows the regression coefficients. The QRS is constructed using at least $p=(m+1)(m+2)/2$ sampled data. The regression model (see equation [5.1]) can be written in matrix form for sampled values of $\boldsymbol{x_S}$ and $\boldsymbol{ys}$, such that:

$$y_S = \boldsymbol{F\beta}$$

with, $\boldsymbol{\beta} = \left[\beta_0, \beta_1, \ldots, \beta_m, \beta_{12}, \ldots, \beta_{(m-1)m}, \beta_{11}, \ldots, \beta_{mm}\right]^T$ is the vector of regression coefficients and $\boldsymbol{y_S} = \left[\hat{y}^{(1)}, \hat{y}^{(2)}, \ldots, \hat{y}^{(n)}\right]^T$ is the response vector.

$$\mathbf{F} = \begin{bmatrix} 1 & x_1^{(1)} & \ldots & x_m^{(1)} & x_1^{(1)}x_2^{(1)} & \ldots & x_{m-1}^{(1)}x_m^{(1)} & (x_1^{(1)})^2 & \ldots & (x_m^{(1)})^2 \\ 1 & x_1^{(2)} & \ldots & x_m^{(2)} & x_1^{(2)}x_2^{(2)} & \ldots & x_{m-1}^{(1)}x_m^{(2)} & (x_1^{(2)})^2 & \ldots & (x_m^{(2)})^2 \\ \vdots & \vdots & \ddots & \vdots & \vdots & \ddots & \vdots & \vdots & \ddots & \vdots \\ 1 & x_1^{(n)} & \ldots & x_m^{(n)} & x_1^{(n)}x_2^{(n)} & \ldots & x_{m-1}^{(1)}x_m^{(n)} & (x_1^{(n)})^2 & \ldots & (x_m^{(n)})^2 \end{bmatrix}$$

$\boldsymbol{F}$ represents the matrix of the model with dimensions $n \times p$ consisting of variable parameters developed according to the shape of the model. In this matrix, each row corresponds to a single design point and each column to a single regression coefficient. The estimators with the least squares for $\boldsymbol{\beta}$ can therefore be expressed as follows:

$$\hat{\boldsymbol{\beta}} = (\boldsymbol{F}^T\boldsymbol{F})^{-1}\boldsymbol{F}^T\boldsymbol{y_S} \qquad [5.2]$$

After calculating the unknown coefficients $\boldsymbol{\beta}$, the approximate response $\hat{y}(\boldsymbol{x})$ can thus be determined by equation [5.1].

The polynomial approximation is both easy to use and implement. However, it has low accuracy when the function to be approximated for certain components is multimodal. When the adjustment range is small, there is enough data available to determine all the regression coefficients in the model, and polynomial metamodels can easily interpolate the adjusted data. However, such small adjustment ranges are generally not recommended. Therefore, these metamodels are mainly used in iterative optimization procedures where a sequence of metamodels is constructed in a progressively smaller region of the design space around the optimal point.

5.2.2.2. *Kriging*

Building surrogate models to reduce the computational cost associated with reliability analysis, stochastic or optimization problems has been given the generic name Design and Analysis for Computer Experiments commonly referred to as DACE in the literature. In this respect, Kriging has emerged over the last two decades as a powerful tool for building metamodels. The basic idea is to model a function known only at a finite number of sampling points in the implementation of a Gaussian random field. Today, Kriging is widely applied due to the robustness and quality of the approximation it can provide. Moreover, it provides an estimate for the least square errors of the process it estimates. The modern Kriging framework (also called Gaussian process modeling [SAR 11]) is based on the fact that the substitution function is assumed to be the result of a Gaussian process denoted $y(x)$. The answer $y(x)$ is formulated as follows:

$$y(\boldsymbol{x}) = G(\boldsymbol{x}) + Z(\boldsymbol{x}) \qquad [5.3]$$

where $G(\boldsymbol{x})$ is usually given as a linear combination of polynomial basis functions:

$$G(\boldsymbol{x}) = \sum_{i=1}^{p-1} \beta_i f_i(\boldsymbol{x}) \qquad [5.4]$$

where $\boldsymbol{f}(\boldsymbol{x}) = [f_0(\boldsymbol{x}), f_1(\boldsymbol{x}), \ldots, f_{p-1}(\boldsymbol{x})]^T$ is the vector of the regression basis function, $\boldsymbol{\beta} = [\beta_0, \beta_1, \ldots, \beta_{p-1}]^T$ is the vector of regression coefficients and $Z(\boldsymbol{x})$ is a stochastic process (random function) with zero mean $\left(\mathbb{E}\left[Z(\boldsymbol{x})\right] = 0\right)$, a variance of σ^2, and non-zero covariance given by:

$$Cov\left(Z(\boldsymbol{x}) - Z(\boldsymbol{x}')\right) = \sigma^2 R(\boldsymbol{x}, \boldsymbol{x}') \qquad [5.5]$$

$R(\boldsymbol{x}, \boldsymbol{x}')$ is the correlation function between the evaluated sample points $\boldsymbol{x}$ and $\boldsymbol{x}'$. This controls the smoothness of the resulting Kriging model and the influence of neighboring points by quantifying the correlation between observations. A product of correlation functions is usually used to construct dimensional functions m:

$$R(\boldsymbol{x} - \boldsymbol{x}') = \prod_{i=1}^{m} R(x_i, x'_i) \qquad [5.6]$$

where m is the dimension of the design variables and x_i and x'_i are, respectively, the ith component of the sample design points $\boldsymbol{x}$ and $\boldsymbol{x}'$.

According to [ELH 13a] and [ELH 13b], two commonly applied functions are exponential functions (see equation [5.8]) and Gaussian functions (see equation [5.9]):

$$R(\boldsymbol{x},\boldsymbol{x}') = \exp\left[-\sum_{i=1}^{m} \theta_i \left|x_i - x'_i\right|\right] \qquad [5.7]$$

$$R(\boldsymbol{x},\boldsymbol{x}') = \exp\left[-\sum_{i=1}^{m} \theta_i \left|x_i - x'_i\right|^2\right] \qquad [5.8]$$

where $\boldsymbol{\theta} = [\theta_1,...,\theta_m]^T$ denotes the vector of unknown parameters of the correlation function. The θ_i parameters are essential parameters that affect the extent to which the influence of a sample point extends [FOR 09]. A low θ_i means that all points will have a high correlation $\boldsymbol{R}$, with $Z(x_i)$ being similar across the sample, while a high means that there is a significant difference between $Z(x_i)$ for different sample points. The Kriging estimator at a given point x is by definition a Gaussian random variable obtained as a linear combination of observations and is the best linear unbiased predictor:

$$\hat{y}(\boldsymbol{x}) = \boldsymbol{f}^T(\boldsymbol{x})\boldsymbol{\beta} + \boldsymbol{r}^T(\boldsymbol{x})\boldsymbol{R}^{-1}\left(\boldsymbol{y_S} - \boldsymbol{F}^T\boldsymbol{\beta}\right) \qquad [5.9]$$

where $\boldsymbol{f}^T(\boldsymbol{x})$ is the vector for the regression basis function of the unknown point x, β denotes the vector for the estimated value of regression coefficients, $\boldsymbol{R}$ is the matrix $m \times m$ of the correlation functions for the adjusted sample $\boldsymbol{r}(\boldsymbol{x}) = [\boldsymbol{R}(\boldsymbol{x},\boldsymbol{x}^{(1)}),\boldsymbol{R}(\boldsymbol{x},\boldsymbol{x}^{(2)}),...,\boldsymbol{R}(\boldsymbol{x},\boldsymbol{x}^{(n)})]^T$ and represents the vector for the correlation functions between the unknown points $\boldsymbol{x}$ and m sample points, $\boldsymbol{y_S}$ represents the vector responses in the adjusted example and $\boldsymbol{F}$ is the model matrix. The term $\left(\boldsymbol{y_S} - \boldsymbol{F}^T\boldsymbol{\beta}\right)$ represents the vector of residuals for all adjusted points when the stochastic model term is neglected. In order to construct the Kriging metamodel, we must first determine the correlation parameters of the vector as well as the variance of the stochastic model σ^2. The method currently most used to determine all these parameters is based on the Kriging statistical framework, and particularly on that of Gaussian processes. Indeed, it proposes to determine the parameters of the model by maximizing the likelihood function that was used to determine the covariance of a random spatial process. The idea is based on the model parameters in such a way that the observation of the responses is as likely as possible, that is to say, such that the probability of the responses is maximized. In the framework of a Universal Kriging, this likelihood is written in the following form:

$$L\left(\beta,\sigma^2,\theta\right)=\frac{1}{\sqrt[n]{2\pi\sigma^2}\sqrt{|R(\theta)|}}\exp\left[-\frac{(y_S-F^T\beta)^T R^{-1}(\theta)\left(y_S-F^T\beta\right)}{2\sigma^2}\right] \quad [5.10]$$

In practice, we apply the Neperian logarithm on the expression of the likelihood, such that:

$$\ln(L)=-\frac{n}{2}\ln(2\pi)-\frac{n}{2}\ln\left(\sigma^2\right)-\frac{1}{2}\ln\left(|R(\theta)|\right)-\frac{(y_S-F^T\beta)^T R^{-1}(\theta)\left(y_S-F^T\beta\right)}{2\sigma^2} \quad [5.11]$$

The likelihood is maximal when its derivative with respect to θ and σ^2 cancels itself, such that:

$$\hat{\boldsymbol{\beta}}=\left(\boldsymbol{F}^T\boldsymbol{R}^{-1}\boldsymbol{F}\right)^{-1}\boldsymbol{F}^T\boldsymbol{R}^{-1}\boldsymbol{y}_S \quad [5.12]$$

and

$$\hat{\sigma}^2=\frac{(y_S-F^T\beta)^T R^{-1}(\theta)\left(y_S-F^T\beta\right)}{n} \quad [5.13]$$

The optimal values of θ are determined by solving the nonlinear optimization problem for maximizing the likelihood function:

$$\begin{aligned}&\max_{\theta} \quad L(\theta)=-\frac{1}{2}\left[n\left(\ln(2\pi)+\ln\left(\hat{\sigma}^2\right)\right)+\ln\left(|R|\right)\right]\\&\text{s.t.} \quad \theta_i>0,\ i=1,..,m\end{aligned} \quad [5.14]$$

Thus, by solving this one-dimensional optimization problem, the estimators $\hat{\beta}$ and $\hat{\sigma}^2$ can be calculated and the model $\hat{y}(\boldsymbol{x})$ is said to be defined by equation [5.10].

5.2.3. *Validation of metamodels*

Once the metamodel is obtained, it is necessary to evaluate its quality. The accuracy of a metamodel is influenced by the type of metamodel and the quality and quantity of the dataset from which it is constructed. The quality of a metamodel cannot be described by a single measure. Instead, there are several methods that can be used to assess the accuracy of a metamodel and compare it to others.

5.2.3.1. *Measurement errors*

One way to assess the accuracy of a metamodel is to look at its residuals, i.e. the difference between the simulated value y_i and the value predicted by the metamodel $\hat{y}_i$. Small residuals mean that the model reflects the data set more accurately than if the residuals were larger, that is to say, the adjustment error is smaller.

The root mean squared error (RMSE) and the relative mean error (RME), are the two most used indicators for measuring the deviation between two functions. These two criteria are used to measure the deviation between the real function and its approximation in a set of n sample points.

$$MAE = \max\left|y^{(i)} - \hat{y}^{(i)}\right|, \qquad i = 1,2,\ldots,n_t \qquad [5.15]$$

$$RME = \frac{1}{n_t}\sum_{i=1}^{n_t}\left|\frac{y^{(i)} - \hat{y}^{(i)}}{y^{(i)}}\right| \qquad [5.16]$$

$$RMSE = \sqrt{\frac{1}{n_t}\sum_{i=1}^{n_t}\left(y^{(i)} - \hat{y}^{(i)}\right)^2} \qquad [5.17]$$

where n_t indicates the number of test points chosen to calculate the error measurements. The smaller these measurement errors are, the more accurate the surrogate model is.

5.2.3.2. *Cross-validation*

Another way to assess the quality of a metamodel and compare it to other models is through cross-validation (CV) [ELH 13a], in which the same dataset is used for the adjustment and validation of the model. When the simulation time is long and the available data is limited, it may be desirable to use the full dataset to adjust the metamodels rather than potentially reducing accuracy by excluding part of the validation set. In *p-fold* CV, the dataset of input-output data for n pairs is divided into different p subsets. The metamodel is then adjusted p times, omitting one of the subsets each time. The omitted subset is used to evaluate the measurement errors of interest. A variation of this method is the *leave-k-out* CV, in which all possible subsets $\binom{n}{k}$ of k sizes are omitted, and the metamodel is adjusted to the remaining set. Each time, the relevant measurement errors are evaluated at the omitted points. This approach is more expensive than the *p-fold* CV. However, for the special case where $k = 1$, called the *leave-one-out* CV, an estimate for the prediction error can be calculated for some metamodels, such as polynomial regression, Kriging and RBF. The generalization error, that is the prediction error, for a *leave-one-out* CV calculation when the error is described by the MSE, can be represented by:

$$MSE_{CV} = \frac{1}{n}\sum_{i=1}^{n}\left(y^{(i)} - \hat{y}_{-i}^{(i)}\right)^2 \qquad [5.18]$$

where $y^{(i)}$ represents the response $x^{(i)}$ using the metamodel constructed from all sample points except $\left(x^{(i)}, y^{(i)}\right)$.

5.3. Optimization of multi-objective design

5.3.1. *Deterministic MOO*

A deterministic MOO design problem consists of minimizing the number of objective functions (cost, mass or volume, etc.) subject to physical, geometrical or functional constraints in the following form:

$$\begin{aligned} \min_{x} \quad & f_m(x), \quad m = 1,..,M \\ \text{s.t.} \quad & g_k(x) \leq 0, \quad k = 1,..,K \\ & h_j(x) = 0, \quad j = 1,..,J \\ & x_{lb} \leq x \leq x_{ub} \end{aligned} \qquad [5.19]$$

where x is the vector of deterministic design variables, x_{lb} and x_{ub} are respectively the lower and upper bounds, $g_k(x)$ are the inequality constraints and $h_j(x)$ denotes the equality constraints.

For the optimization of multi-objective design, the objective functions constitute a multi-dimensional space called the *objective space*, in addition to the design variable space when compared to standard single-objective optimization. Figure 5.2 shows a transformation of design variables from the physical space to the objective function space.

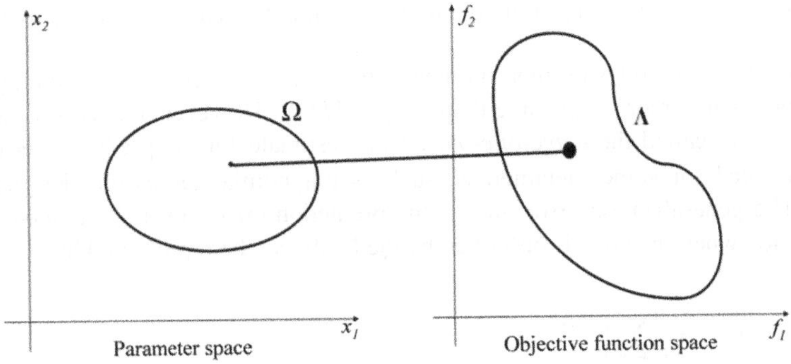

Figure 5.2. *Transformation from a parameter space to an objective function space*

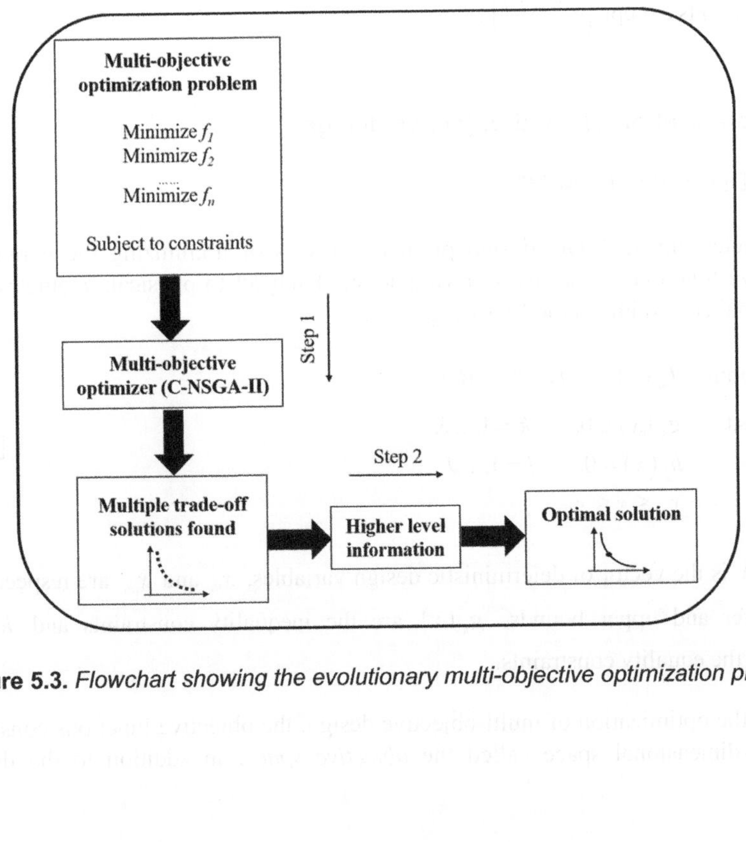

Figure 5.3. *Flowchart showing the evolutionary multi-objective optimization process*

The objective of solving a multi-objective design optimization is to obtain the set of optimal solutions for the Pareto front. After that, the designer can use higher-level qualitative considerations to choose between these solutions. Since EMO approaches deal with a population of solutions at each iteration, this makes them intuitive to apply when seeking a set of non-dominated solutions [DEB 02]. An EMO-based procedure when handling MOO problems is based on two steps, as shown in Figure 5.3.

– step 1: find several non-dominated points as close as possible to the Pareto optimal front, with a broad compromise between objectives;

– step 2: choose the best point based on qualitative information.

5.3.2. *Reliability-based multi-objective design optimization*

Different from deterministic design optimization, RBDO seeks the best trade-off between cost and safety when uncertainty in design variables and parameters is taken into account, thereby ensuring that the desired performance is achieved with a certain probability. An RBMDO problem can therefore be formulated as follows:

$$\begin{aligned} \min_{x} \quad & f_m(x), \quad m = 1,..,M \\ \text{s.t.} \quad & P_r\left[G_k(x,y) \le 0\right] \le P_k^T, \quad k = 1,..,K \\ & h_j(x) \le 0, \quad j = 1,..,J \end{aligned} \qquad [5.20]$$

where x is the vector of deterministic design variables, y is the vector of random variables, $P_r[.]$ represents the probability operator, P_i^T indicates the target probability of failure, $G_i(x,y)$ are the probabilistic constraints and $h_j(x)$ are the deterministic constraints.

In general, the RBDO procedure is solved by the classical approach in two spaces: the physical space of design variables and the normalized space of random variables. It has been proven that this approach requires an extended computational time due to a large number of numerical calculations in both spaces. Therefore, a more efficient approach, called the HM [KHA 08, YAI 18, DAM 20], is needed to reduce the effects of this problem. The basic idea of this approach, in the context of the single-objective function, is to minimize a new form of the objective function subject to the performance function $G(x, y)$, the deterministic constraints $g_k(x)$ and the reliability constraint. The transformation of this method in the context of multi-objective functions leads to the optimization problem being described as follows:

$$\begin{aligned} \min_{x,y} \quad & F_m(x,y) = f_m(x).d_\beta(x,y), \qquad m = 1,..,M \\ \text{s.t.} \quad & G(x,y) \le 0 \\ & g_k(x,y) \le 0, \qquad k = 1,..,K \\ & d_\beta(x,y) \ge \beta_t \end{aligned} \qquad [5.21]$$

where $d_\beta(x,y)$ indicates the distance between the optimal solutions and the design points and β_t is the target reliability. The optimization problem is solved using the multi-objective genetic algorithm and the Non-dominated Sorting Genetic Algorithm (NSGA-II) developed by Deb et al. [DEB 02].

5.4. RBMDO based on hip prosthesis surrogate models

In the proposed method, deterministic finite element simulation (FES) is used to obtain the characteristic response at the generated points using LHS, next, the metamodel is used to illustrate the relationship between the inputs (design points) and outputs (responses), and finally, the multi-objective reliability optimization is performed using the constructed surrogate model [DAM 20].

5.4.1. *Deterministic simulation using the finite element method*

It is well known that a bone must adapt to the applied load. Bone materials and their structures are also complicated. According to Mackerle [MAC 06], bone is considered an anisotropic material with 21 parameters for determining its stiffness. For simplicity, the material properties of bone are assumed to be linearly elastic and isotropic. Figure 5.4 shows a 3D model of the solid shaft being studied and a corresponding 2D model. The simplified geometry is composed of different layers.

The FES was performed using ANSYS software. The solid shaft was modeled with 1,447 elements (eight nodes/PLANE82), giving a total of 4,736 nodes. According to [KHA 15], the bone marrow layer can be treated as a cavity, so it is ignored in the FES. Table 5.1 shows the properties of the model being studied.

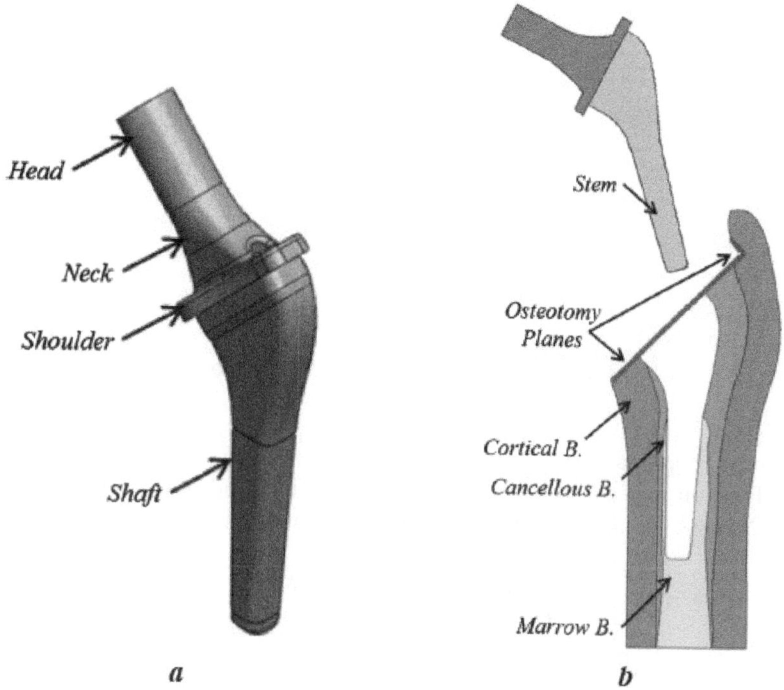

Figure 5.4. *Diagram showing a) 3D shaft model and b) 2D shaft with different layers*

	Elasticity (MPa)	Poisson coefficient	Stress rupture (MPa)	Number of elements
Shaft	110,000	0.3	825	534
Cortical layer	17,000	0.3	132	699
Cancellous layer	761	0.3	10.53	212

Table 5.1. *Properties of the model studied*

According to [KHA 09], the most critical loading case for adduction is presented in Figure 5.5.

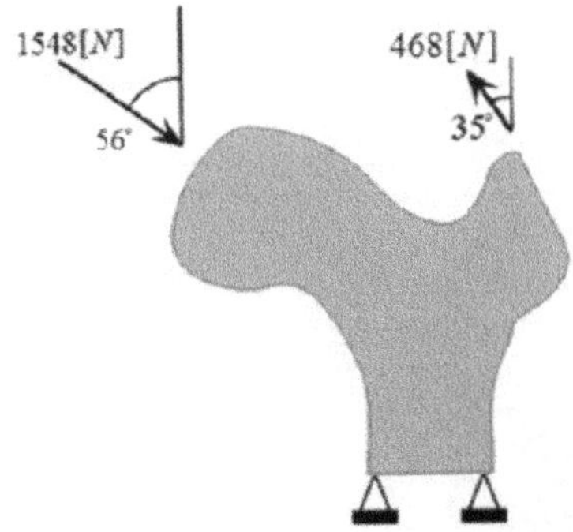

Figure 5.5. *Adduction boundary conditions*

Boundary conditions at the distal end have no effect on stresses in the proximal region. Fixation is performed on the cortical layer to avoid rigid body movement. Table 5.2 shows the maximum values of von-Mises stresses for the different layers.

Variables	Values (MPa)
Cortical stress	15.59
Cancellous stress	5.63
Shaft stress	11.28

Table 5.2. *Maximum von-Mises stresses of the different layers*

5.4.2. *Construction of substitution models*

The design variables of the optimization problem correspond to Young's modulus, the Poisson's ratio of the different layers, and the dimensions of the shaft, which are presented in Table 5.3.

Variables	Average values	Covariance	Distribution
Cortical elasticity (MPa)	110,000	0.1	Normal
Cortical elasticity (MPa)	17,000	0.1	Normal
Cortical elasticity (MPa)	761	0.1	Normal
Cortical Poisson coefficient	0.3	0.02	Normal
Cancellous Poisson coefficient	0.3	0.02	Normal
Shaft Poisson coefficient	0.3	0.02	Normal

Table 5.3. *Characteristics of design variables*

The maximum von-Mises stresses for the different layers are approximated using the QRS (see equation [5.1]) and Kriging (see equation [5.10]) metamodels. The Kriging models are obtained using the MATLAB toolbox: DACE [ABI 20]. The second-order global polynomial trend function and the exponential correlation function are adopted. The estimators are constructed through 100 LHS points. An additional 100 sample points are generated to validate the accuracy of the proxy models. The FESs run multiple times using the training data (see Table 5.3) as inputs. The 100 FES runs using LHS points are implemented in the MATLAB and ANSYS interoperation. After generating the sample data, the surrogate model can be constructed. The accuracy of both types of metamodels was defined by their MAE (see equation [5.15]), RME (see equation [5.17]), and RMSE (see equation [5.18]). The results of the error analysis of the surrogate models are presented in Table 5.4.

Measurement errors	Cortical stress		Cancellous stress		Shaft stress	
	QRS	Kriging	QRS	Kriging	QRS	Kriging
MAE	3.814	2.104	0.097	0.092	0.3	0.302
RME	0.034	0.033	0.005	0.006	0.008	0.007
RMSE	0.09	0.008	0.005	0.003	0.004	0.003

Table 5.4. *Comparing the accuracy of substitution models*

A more significant result for the different layers is shown from CV, as seen in Figures 5.6–5.8.

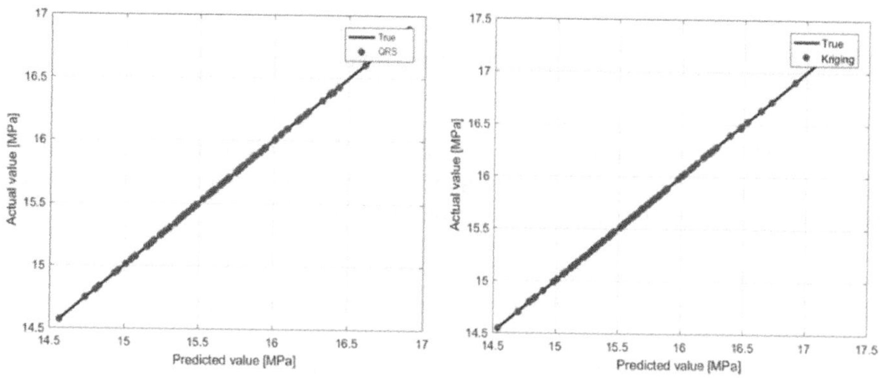

Figure 5.6. *Cross-validation for maximum cortical layer stress with a) QRS and b) Kriging*

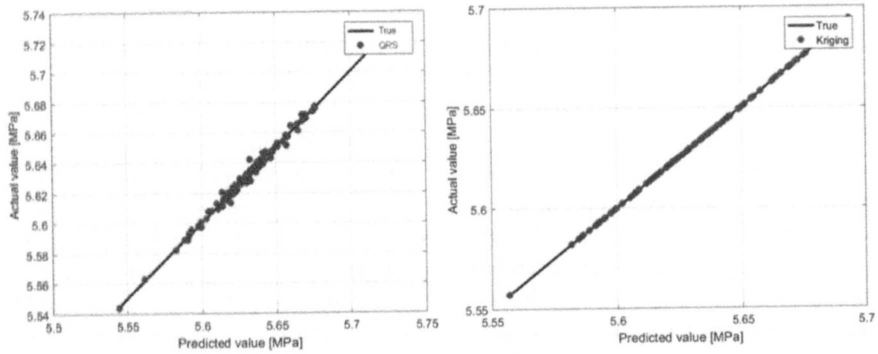

Figure 5.7. *Cross-validation for the maximum stress of the cancellous layer with a) QRS and b) Kriging*

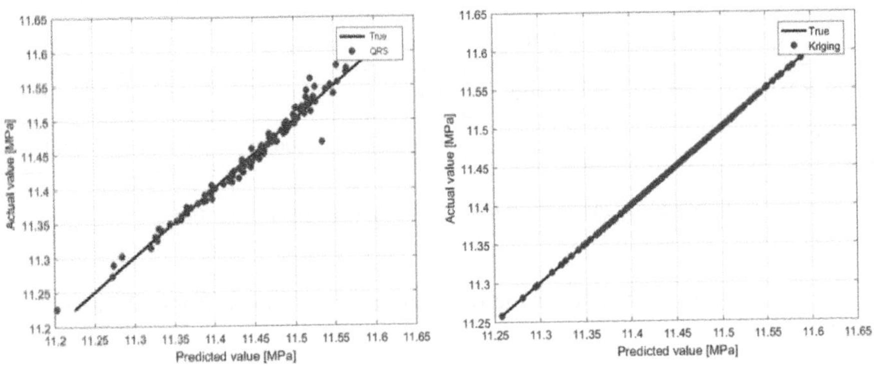

Figure 5.8. *Cross-validation for maximum shaft stress with a) QRS and b) Kriging*

From Table 5.4 and the CV figures, we can see that the Kriging models are more accurate than the QRS models for approximating the hip replacement model. Therefore, it will be used in the optimization problem.

5.4.3. *Optimization of multi-objective design based on reliability*

The optimization problem is solved using the constrained NSGA-II (C-NSGA-II), developed by Deb et al. [DEB 02].

The optimization problem is to minimize the difference between the maximum and minimum stress values for the cortical and cancellous layers subjected to stress ruptures. Here, the reliability constraint can be defined using the HM, by converting the deterministic stresses into reliability stresses ($\beta_t = 3$), as follows:

$$\begin{array}{ll} \min\limits_{x,y} & F(x,y) = [f_1(x), f_2(x)].d_\beta(x,y) \\ \text{s.t.} & \sigma_{max}^{Cor}(x,y) \le \sigma_y^{Cor} \\ & \sigma_{max}^{Can}(x,y) \le \sigma_y^{Can} \\ & \sigma_{max}^{Met}(x,y) \le \sigma_y^{Met} \\ & d_\beta(x,y) \ge \beta_t \\ & lb \le x, y \le ub \end{array} \quad [5.22]$$

where $f_1(x) = \sigma_{max}^{Cor}(x,y) - \sigma_{min}^{Cor}(x,y)$, $f_2(x) = \sigma_{max}^{Can}(x,y) - \sigma_{min}^{Can}(x,y)$, and lb , are the lower and upper values of the respective design variables.

In this application, we use a population size of 200 with a maximum of 500 generations. Figures 5.9 and 5.10 show the Pareto optimal front of DMOO and RBMDO, respectively. The minimum distance selection method (MDSM) is used here to select the best optimum. This particular point on the Pareto front generally represents a good compromise between the objectives.

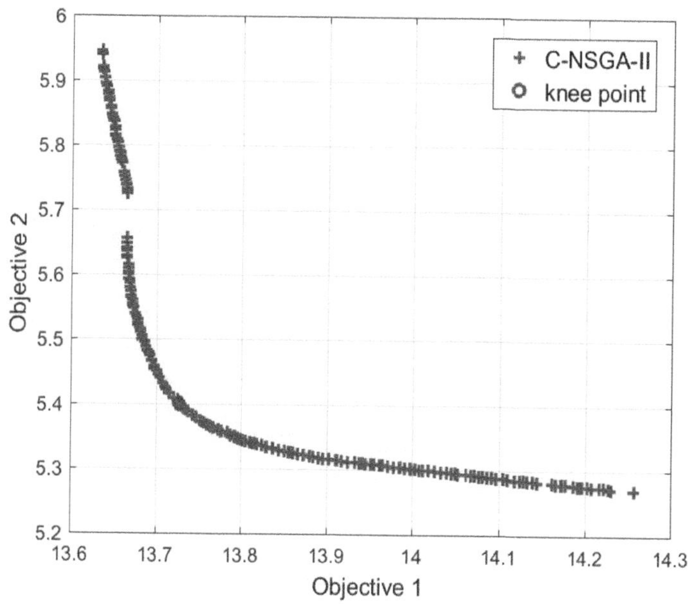

Figure 5.9. *DMMO Pareto optimal front*

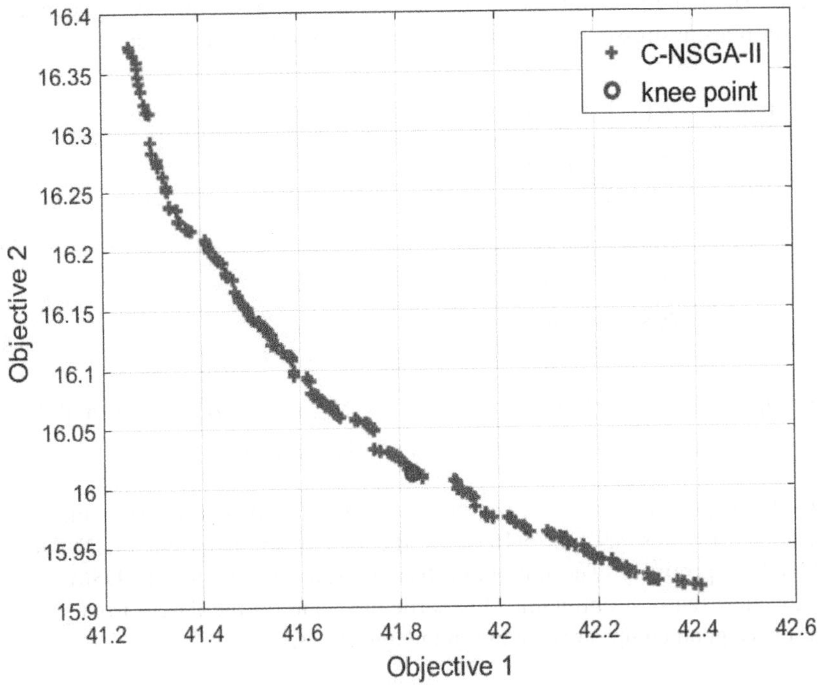

Figure 5.10. *RBMDO Pareto optimal front*

Like the DMOO, RBMDO presents a set of solutions in the Pareto space. However, because the design variables are related to uncertainties, the Pareto RBMDO front differs considerably from that for DMOO. A comparison between the initial design and the two optimization problems (DMOO and RBMDO) is presented in Table 5.5.

Compared to the baseline design, it can be seen that although deterministic optimization generated the best hip prosthesis design, the reliability of this design was the worst. Therefore, a trade-off between the desired level of reliability and the objective performance must be found. With respect to the level of reliability, a considerable improvement in the RBMDO design was achieved. The distribution of the von-Mises stresses for different configurations is shown in Figure 5.11.

Parameters	Initial model	DMOO	RBMDO
Cortical elasticity (MPa)	11,000	109,590.37	108,488.03
Cancellous elasticity (MPa)	17,000	18,351.51	18,645.9
Stress elasticity (MPa)	761	835.1	835.08
Cortical Poisson coefficient	0.3	0.359	0.328
Cancellous Poisson coefficient	0.3	0.3	0.318
Shaft Poisson coefficient	0.3	0.301	0.302
Objective 1 (MPa)	13.81	13.73	13.94
Objective 2 (MPa)	5.21	5.15	5.14
Cortical stress (MPa)	15.59	15.39	15.49
Cancellous stress (MPa)	5.63	6.29	6.32
Shaft stress (MPa)	11.45	11.39	11.41
Reliability index	1.29	1.78	3
Volume (mm^3)	5,675.22	5,625.94	5,618.08

Table 5.5. *Comparison between the initial model, DMOO and RBMDO models*

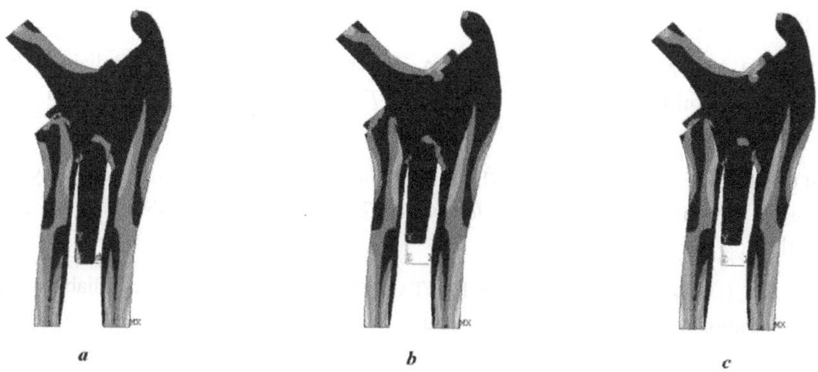

Figure 5.11. *von-Mises distribution: a) Initial b) DMOO and c) RBMDO*

5.5. Conclusion

This chapter aims to present a methodology to design a cementless hip prosthesis with optimal performance. The proposed methodology consists of a combination of

FES, substitution techniques, and optimization procedures to reduce the computational cost. The reliable and optimized design was obtained by considering the uncertainties of the design variables. In order to find a compromise for the difference between the maximum and minimum stress values of cortical and spongy layers, and the yield strength values for all layers, the DMOO and RBMDO procedures were adopted. The C-NSGA-II is used to generate Pareto optimal fronts. The MDSM is performed to find the best point of the Pareto front with respect to the reliability constraint. The proposed RBMDO using the Kriging metamodel is able to generate reliable and well-distributed Pareto solutions. The results also show that with RBMDO, a great improvement in the design with respect to the reliability level was achieved.

5.6. References

[ABI 20] ABID F., DAMMAK K., EL HAMI A., et al., "Surrogate models for uncertainty analysis of micro-actuator", *Microsystem Technologies*, vol. 26, pp. 2589–2600, 2020.

[ABO 11a] ABO AL-KHEER A., EID M., AOUES Y., et al., "Theoretical analysis of the spatial variability in tillage forces for fatigue analysis of tillage machines", *Journal of Terramechanics*, vol. 48, no. 4, pp. 285–295, 2011.

[ABO 11b] ABO AL-KHEER A., KHARMANDA M.G., EL HAMI A., et al., "Estimating the variability of tillage forces on a chisel plough shank by modeling the variability of tillage system parameters", *Computers and Electronics in Agriculture*, vol. 78, no. 1, pp. 61–70, 2011.

[ABO 11c] ABO AL-KHEER A., EL HAMI A., KHARMANDA M.G., et al., "Reliability-based design for soil tillage machines", *Journal of Terramechanics*, vol. 48, no. 1, pp. 57–64, 2011.

[ABO 11d] ABO AL-KHEER A., EL HAMI A., KHARMANDA M.G., et al., "Reliability-based design for soil tillage machines", *Journal of Terramechanics*, vol. 48, no. 1, pp. 57–64, 2011.

[BAR 04] BARAKAT S., BANI-HANI K., TAHA M.Q., "Multi-objective reliability-based optimization of prestressed concrete beams", *Structural Safety*, vol. 26, no. 3, pp. 311–342, 2004.

[BEN 13] BEN ABDESSALEM A., PAGNACCO E., EL HAMI A., "Increasing the stability of T-shape tube hydroforming process under stochastic framework", *International Journal of Advanced Manufacturing Technology*, vol. 69, nos 5–8, pp. 1343–1357, 2013.

[BEN 15] BEN ABDESSALEM A., EL HAMI A., "A probabilistic approach for optimising hydroformed structures using local surrogate models to control failures", *International Journal of Mechanical Sciences*, vols 96–97, pp. 143–162, 2015.

[BOU 96] BOUYEKHF R., EL MOUDNI A., EL HAMI A., et al., "Reduced order modelling of two-time-scale discrete non-linear systems", *Journal of the Franklin Institute*, vol. 333, no. 4, pp. 499–512, 1996.

[BOU 01] BOUYEKHF R., EL HAMI A., EL MOUDNI A., "Optimal control of a particular class of singularly perturbed nonlinear discrete-time systems", *IEEE Transactions on Automatic Control*, vol. 46, no. 7, pp. 1097–1101, 2001.

[BOU 09] BOUAZIZI M.L., GHANMI S., BOUHADDI N., "Multi-objective optimization in dynamics of the structures with nonlinear behavior: Contributions of the metamodels", *Finite Elements in Analysis and Design*, vol. 45, no. 10, pp. 612–623, 2009.

[CHO 12] CHO J.R., LEE J.H., JEONG K.M., et al., "Optimum design of run-flat tire insert rubber by genetic algorithm", *Finite Elements in Analysis and Design*, vol. 52, pp. 60–70, 2012.

[CHR 08] CHEROUAT A., RADI B., EL HAMI A., "The frictional contact of the composite fabric's shaping", *Acta Mechanica*, vol. 199, nos 1–4, pp. 29–41, 2008.

[DAM 17] DAMMAK K., EL HAMI A., KOUBAA S., et al., "Reliability based design optimization of coupled acoustic-structure system using Generalized Polynomial Chaos", *International Journal of Mechanical Sciences*, vol. 134, pp. 75–84, 2017.

[DAM 19a] DAMMAK K., EL HAMI A., "Multi-objective reliability-based design optimization of coupled acoustic-structure system", *Engineering Structures*, vol. 197, 109389, 2019.

[DAM 19b] DAMMAK K., KOUBAA S., EL HAMI A., et al., "Numerical modelling of vibro-acoustic problem in presence of uncertainty: Application to a vehicle cabin", *Applied Acoustics*, vol. 144, pp. 113–123, 2019.

[DAM 20a] DAMMAK K., YAICH A., EL HAMI A., et al., "An efficient optimization based on the robust hybrid method for the coupled acoustic–structural system", *Mechanics of Advanced Materials and Structures*, vol. 27, no. 21, pp. 1816–1826, 2020.

[DAM 20b] DAMMAK K., EL HAMI A., "Multi-objective reliability-based design optimization using Kriging surrogate model for cementless hip prosthesis", *Computer Methods in Biomechanics and Biomedical Engineering*, vol. 23, no. 12, pp. 854–865, 2020.

[DAM 21] DAMMAK K., El HAMI A., "Thermal reliability-based design optimization using Kriging model of PCM based pin fin heat sink", *International Journal of Heat and Mass Transfer*, vol. 166, 120745, 2021.

[DAS 11] DASKILEWICZ M.J., GERMAN B.J., TAKAHASHI T.T., et al., "Effects of disciplinary uncertainty on multi-objective optimization in aircraft conceptual design", *Structural and Multidisciplinary Optimization*, vol. 44, no. 6, pp. 831–846, 2011.

[DEB 02] DEB K., PRATAP A., AGARWAL S., et al., "A fast and elitist multiobjective genetic algorithm: NSGA-II", *IEEE Transactions on Evolutionary Computation*, vol. 6, no. 2, pp. 182–195, 2002.

[DEB 07] DEB K., PADMANABHAN D., GUPTA S., et al., "Reliability-based multi-objective optimization using evolutionary algorithms", in OBAYASHI S., DEB K., POLONI C., et al. (eds), *International Conference on Evolutionary Multi-Criterion Optimization*, Springer, Berlin, Heidelberg, 2007.

[DEB 11] DEB K., "Multi-objective optimization using evolutionary algorithms: An introduction", in WANG L., NG A.H.-C., DEB K. (eds), *Multi-objective Evolutionary Optimization for Product Design and Manufacturing*, Springer-Verlag London Limited, London, DOI:10.1007/978-0-85729-652-8_1, 2011.

[DEB 21] DEBICH B., Modélisation, simulation, et fiabilité des matériaux changement de phase pour l'automobile et l'aéronautique, Thesis, INSA Rouen Normandie, 2021.

[DEB 22] DEBICH B., EL HAMI A., YAICH A., et al., "An efficient reliability-based design optimization study for PCM-based heat-sink used for cooling electronic devices", *Mechanics of Advanced Materials and Structures*, vol. 29, no. 12, pp. 1661–1673, 2022.

[DEY 15] DEY S., MUKHOPADHYAY T., KHODAPARAST H.H., et al., "Rotational and ply-level uncertainty in response of composite shallow conical shells", *Composite Structures*, vol. 131, pp. 594–605, 2015.

[DEY 17] DEY S., MUKHOPADHYAY T., ADHIKARI S., "Metamodel based high-fidelity stochastic analysis of composite laminates: A concise review with critical comparative assessment", *Composite Structures*, vol. 171, pp. 227–250, 2017.

[ELH 93] EL HAMI A., LALLEMENT G., MINOTTI P., et al., "Methods that combine finite group theory with component mode synthesis in the analysis of repetitive structures", *Computers and Structures*, vol. 48, no. 6, pp. 975–982, 1993.

[ELH 96] EL HAMI A., RADI B., "Some decomposition methods in the analysis of repetitive structures", *Computers and Structures*, vol. 58, no. 5, pp. 973–980, 1996.

[ELH 09] EL HAMI A., RADI B., CHEROUAT A., "Treatment of the composite fabric's shaping using a Lagrangian formulation", *Mathematical and Computer Modelling*, vol. 49, nos 7–8, pp. 1337–1349, 2009.

[ELH 11] EL HAMI A., RADI B., "Comparison study of different Reliability-Based Design Optimization approaches", *Advanced Materials Research*, vol. 274, pp. 113–121, 2011.

[ELH 13a] EL HAMI A., RADI B., *Incertitude et optimisation et fiabilité des structures*, Hermes-Lavoisier, Paris, 2013.

[ELH 13b] EL HAMI A., RADI B., *Uncertainty and Optimization in Structural Mechanics*, ISTE Ltd, London and John Wiley & Sons, New York, 2013.

[FAN 15] FANG J., GAO Y., SUN G. et al., "Multi-objective robust design optimization of fatigue life for a truck cab", *Reliability Engineering and System Safety*, vol. 135, pp. 1–8, 2015.

[FOR 08] FORRESTER A., SOBESTER A., KEANE A., *Engineering Design via Surrogate Modelling: A Practical Guide*, John Wiley & Sons, New York, 2008.

[FOR 09] FORRESTER A.I., KEANE, A.J., "Recent advances in surrogate-based optimization", *Progress in Aerospace Sciences*, vol. 45, nos 1–3, pp. 50–79, 2009.

[GOE 07] GOEL T., VAIDYANATHAN R., HAFTKA R.T., et al., "Response surface approximation of Pareto optimal front in multi-objective optimization", *Computer Methods in Applied Mechanics and Engineering*, vol. 196, nos 4–6, pp. 879–893, 2007.

[GUE 15a] GUERINE A., EL HAMI A., WALHA L., et al., "Approach for the dynamic analysis of one stage gear system with uncertain parameters", *Mechanism and Machine Theory*, vol. 92, pp. 113–126, 2015.

[GUE 15b] GUERINE A., EL HAMI A., FAKHFAKH T., et al., "A polynomial chaos method to the analysis of the dynamic behavior of spur gear system", *Structural Engineering and Mechanics*, vol. 53, no. 4, pp. 819–831, 2015.

[GUE 16a] GUERINE A., EL HAMI A., WALHA L., et al., "A polynomial chaos method for the analysis of the dynamic behavior of uncertain gear friction system", *European Journal of Mechanics A/Solids*, vol. 59, pp. 76–84, 2016.

[GUE 16b] GUERINE A., EL HAMI A., WALHA L., et al., "Dynamic response of a Spur gear system with uncertain friction coefficient", *Advances in Engineering Software*, vol. 120, pp. 45–54, 2016.

[GUE 17] GUERINE A., EL HAMI A., WALHA L., et al., "Dynamic response of wind turbine gear system with uncertain-but-bounded parameters using interval analysis method", *Renewable Energy*, vol. 113, pp. 679–687, 2017.

[HUA 16] HUANG C., RADI B., EL HAMI A., "Uncertainty analysis of deep drawing using surrogate model based probabilistic method", *Manufacturing Technology*, vol. 86, nos 9–12, pp. 3229–3240, 2016.

[JAN 17] JANNOUN M., AOUES Y., PAGNACCO E., et al., "Probabilistic fatigue damage estimation of embedded electronic solder joints under random vibration", *Microelectronics Reliability*, vol. 78, pp. 249–257, 2017.

[KAM 22] KAMEL A., DAMMAK K., EL HAMI A., et al., "A modified hybrid method for a reliability-based design optimization applied to an offshore wind turbine", *Mechanics of Advanced Materials and Structures*, vol. 29, no. 9, pp. 1229–1242, 2022.

[KHA 08] KHARMANDA G., MOHSINE A., MAKLOUFI A., et al., "Recent methodologies for reliability-based design optimization", *International Journal for Simulation and Multidisciplinary Design Optimization*, vol. 2, no. 1, pp. 11–23, 2008.

[KHA 09] KHARMANDA G., SHARABATEY S., IBRAHIM H., et al., "Reliability-based design optimization using semi-numerical strategies for structural engineering applications", *International Journal of CAD/CAM*, vol. 9, no. 1, pp. 1–16, 2009.

[KUR 02] KURTARAN H., ESKANDARIAN A., MARZOUGUI D., et al., "Crashworthiness design optimization using successive response surface approximations", *Computational Mechanics*, vol. 29, no. 4, pp. 409–421, 2002.

[LAU 13] LAURENT L., BOUCARD P.A., SOULIER B., "Generation of a cokriging metamodel using a multiparametric strategy", *Computational Mechanics*, vol. 51, no. 2, pp. 151–169, 2013.

[LED 17a] LEDOUX M., El HAMI A., *Fluid Mechanics: Analytical Methods*, ISTE Ltd, London and John Wiley & Sons, New York, 2017.

[LED 17b] LEDOUX M., El HAMI A., *Compressible Flow, Propulsion and Digital Approaches in Fluid Dynamics*, ISTE Ltd, London and John Wiley & Sons, New York, 2017.

[LED 21a] LEDOUX M., EL HAMI A., *Heat Transfer 1. Conduction*, ISTE Ltd, London and John Wiley & Sons, New York, 2021.

[LED 21b] LEDOUX M., EL HAMI A., *Heat Transfer 2. Radiative Transfer*, ISTE Ltd, London and John Wiley & Sons, New York, 2021.

[LI 19] LI Z., SU X., TAN J., et al., "Multi-objective optimization of the layout of damping material for reducing the structure-borne noise of thin-walled structures", *Thin-Walled Structures*, vol. 140, pp. 331–341, 2019.

[LIA 08] LIAO X., LI Q., YANG X., et al., "Multiobjective optimization for crash safety design of vehicles using stepwise regression model", *Structural and Multidisciplinary Optimization*, vol. 35, no. 6, pp. 561–569, 2008.

[MAB 17] MABROUK I.B., EL HAMI A., WALHA L., et al., "Dynamic vibrations in wind energy systems: Application to vertical axis wind turbine", *Mechanical Systems and Signal Processing*, vol. 85, pp. 396–414, 2017.

[MAC 06] MACKERLE J., "Finite element modeling and simulations in orthopedics: A bibliography 1998–2005", *Computer Methods in Biomechanics and Biomedical Engineering*, vol. 9, no. 3, pp. 149–199, 2006.

[MAK 16] MAKHLOUFI A., AOUES Y., EL HAMI A., "Reliability based design optimization of wire bonding in power microelectronic devices", *Microsystem Technologies*, vol. 22, no. 12, pp. 2737–2748, 2016.

[MCK 00] MCKAY M.D., BECKMAN R.J., CONOVER W.J., "A comparison of three methods for selecting values of input variables in the analysis of output from a computer code", *Technometrics*, vol. 42, no. 1, pp. 55–61, 2000.

[MEC 01] MECKESHEIMER M., BARTON R.R., SIMPSON T.W., et al., "Computationally inexpensive metamodel assessment strategies", *International Design Engineering Technical Conferences and Computers and Information in Engineering Conference*, American Society of Mechanical Engineers, vol. 80227, pp. 191–201, 2001.

[MOH 10] MOHSINE A., EL HAMI A., "A robust study of reliability-based optimization methods under eigen-frequency", *Computer Methods in Applied Mechanics and Engineering*, vol. 199, nos 17–20, pp. 1006–1018, 2010.

[MOR 02] MORO T., EL HAMI A., EL MOUDNI A., "Reliability analysis of a mechanical contact between deformable solids", *Probabilistic Engineering Mechanics*, vol. 17, no. 3, pp. 227–232, 2002.

[MUK 17] MUKHOPADHYAY T., CHAKRABORTY S., DEY S., et al., "A critical assessment of Kriging model variants for high-fidelity uncertainty quantification in dynamics of composite shells", *Archives of Computational Methods in Engineering*, vol. 24, no. 3, pp. 495–518, 2017.

[MYE 16] MYERS R.H., MONTGOMERY D.C., ANDERSON-COOK C.M., *Response Surface Methodology: Process and Pshaftuct Optimization using Designed Experiments*, John Wiley & Sons, New York, 2016.

[RAD 07] RADI B., EL HAMI A., "Reliability analysis of the metal forming process", *Mathematical and Computer Modelling*, vol. 45, nos 3–4, pp. 431–439, 2007.

[SAN 03] SANTNER T.J., WILLIAMS B.J., NOTZ W.I., et al., *The Design and Analysis of Computer Experiments*, Springer, New York, 2003.

[SAR 11] SARSRI D., AZRAR L., JEBBOURI A., et al., "Component mode synthesis and polynomial chaos expansions for stochastic frequency functions of large linear FE models", *Computers and Structures*, vol. 89, nos 3–4, pp. 346–356, 2011.

[SIM 01] SIMPSON T.W., POPLINSKI J.D., KOCH P.N., et al., "Metamodels for computer-based engineering design: Survey and recommendations", *Engineering with Computers*, vol. 17, no. 2, pp. 129–150, 2001.

[WAN 17] WANG Y., QIN X., HUANG S., et al., "Structural-borne acoustics analysis and multi-objective optimization by using panel acoustic participation and response surface methodology", *Applied Acoustics*, vol. 116, pp. 139–151, 2017.

[YAI 18] YAICH A., KHARMANDA G., EL HAMI A., et al., "Reliability based design optimization for multiaxial fatigue damage analysis using robust hybrid method", *Journal of Mechanics*, vol. 34, no. 5, pp. 551–566, 2018.

[ZHA 18] ZHANG J., XIAO M., GAO L., et al., "A novel projection outline based active learning method and its combination with Kriging metamodel for hybrid reliability analysis with random and interval variables", *Computer Methods in Applied Mechanics and Engineering*, vol. 341, pp. 32–52, 2018.

6

CMA-ES Assisted by the Kriging Metamodel for the Optimization of Thermomechanical Performances of Mechatronic Packaging

Failures in mechatronic systems are often caused by the fatigue failure of solder joints of their electronic components. With the increasing miniaturization of electronic packages, the stress on solder joints, which ensure the connection between the component outputs and the board, can become problematic. Although solder joints can handle high deformation rates, an accumulation of repeated solicitations instigates their premature aging, which in turn can lead to their failure (thermomechanical fatigue phenomenon). Therefore, studies based on finite element method simulations are conducted to numerically study the lifetime of the packages mounted on the printed circuit board (second level reliability). The high computational cost of solving this type of problem may make the optimization and reliability evaluation procedure unfeasible, due to the high computational time of the multiphysics finite element method simulations. In this chapter, an adaptation of the CMA-ES (covariance matrix adaption evolution strategy) method assisted by the Kriging metamodel is proposed for the global optimization of a problem. The implementation of this proposed algorithm has been carried out in the global optimization of solder joints for plastic quad flat package (PQFP) housing. The results of numerical studies show that the proposed Kriging-assisted covariance matrix adaptation evolution strategy (KA-CMA-ES) algorithm is more efficient and effective than the standard CMA-ES algorithm.

Chapter written by Hamid HAMDANI, Bouchaib RADI and Abdelkhalak EL HAMI.

For a color version of all the figures in this chapter, see: http://www.iste.co.uk/elhami/applied1.zip.

6.1. Introduction

The rapid growth of mechatronic systems, with ever smaller devices, makes the reliability of these systems a major challenge. Most failures of mechatronic devices are due to solder joint fatigue of the electronic components. However, to meet the reliability requirements, the component must be reliably optimized, the robustness of the solder joints must be evaluated and their connections must be validated. Thermal cycling tests are the most commonly used to assess the thermomechanical reliability of solder joints. The purpose of these tests is to investigate the ability of solder joints to withstand mechanical stresses, which are a major failure mechanism that can lead to fatigue failure of solder joints [MAK 16, SAR 11, ELH 13a]. Thermomechanical stresses, induced by alternating thermal cycles, can lead to plastic deformation, the accumulation of which can cause failure of the solder joint and consequently failure of the entire system. To avoid operational failures and reduce warranty costs, robust optimization and validation of critical joints must be performed in the design and manufacture of mechatronic devices. The robustness of this validation includes optimizing the design to maximize the number of fatigue lifecycles of the device in question.

Generally, in structural design optimization or reliability-based design optimization (RBDO) [ELH 13a], derivative-based algorithms are the most widely used methods. These methods require a derivative calculation. Compared to other methods, the main advantage of these methods is that they require a much smaller number of iterations to converge to an optimum. Nevertheless, only the convergence to a local minimum is guaranteed. However, derivative-free algorithms, which are based only on the objective function, have proven to be powerful tools for global search for *optima* and are therefore widely used in solving real-world problems such as engineering optimization. Evolution strategies (ESs) are powerful derivative-free algorithms that are popular for black-box optimization, where no so-called objective function expression is known and no derivatives can be computed [KRA 14]. The use of ESs in real applications is a testament to their capability [LAC 13]. However, the drawback of ESs is the large number of evaluations of the objective function needed to obtain a satisfactory solution. Moreover, in engineering applications, such as mechatronic systems, objective function evaluations are performed using expensive means such as numerical methods (finite element method). As such, the computational cost produced by a large number of objective function evaluations remains the main challenge in the application of ESs.

Metamodel-assisted evolution strategies (MAES) have been developed to overcome expensive numerical simulations using the finite element method. The metamodel is designed to approximate and be used with the original expensive function (finite element code). Several techniques for MAES have been presented by several researchers [JIN 05]. Specifically, MAES can be used with several types of metamodels to approximate the original objective function [HAM 17], such as the Kriging metamodel, radial basis functions (RBFs), support vector machines (SVMs), the K-nearest neighbor (KNN) method and artificial neural networks (ANNs). However, metamodel management and selection, which mainly depend on the problem to be addressed, remain the two main issues in the development and application of MAES.

In this chapter, the KA-CMA-ES algorithm is used to optimize the thermomechanical performance of solder joints in a PQFP package. The first part consists of developing a numerical model to simulate the thermomechanical behavior of the solder joints [ELH 13a]. The finite element analysis software ANSYS is the tool used to model the system. The viscoplastic behavior of the joints is considered using the Anand model, and the results of the inelastic deformation due to temperature variations are presented. In the second part, the main features of the KA-CMA-ES method are presented, and then a formulation of the KA-CMA-ES algorithm is described to optimize the thermomechanical performance of solder joints by maximizing the number of lifecycles of a PQFP-type package [ELH 13a].

6.2. Presentation of the system under study

In this study, the system analyzed is a plastic microcontroller of the “plastic quad flat package” (PQFP) variety; it is a large component of dimensions (40 mm × 28 mm) and a very fine pin pitch (0.5 mm) (see Figure 6.2). This type of component is implemented in several nomenclatures, such as that of engine control units in automobiles. The risk for this component during the life of the mechatronic device is the opening of the solder joints electrically connecting the pins of the component to the printed circuit board. The stress triggering the failure mechanism is the accumulation of temperature variation cycles. The failure mechanism is the fatigue cracking of the solder joints caused by thermomechanical stress cycles in the solder joint material and its interfaces.

Figure 6.1 shows the location of the PQFP microcontroller in the mechatronic (automotive) system [ELH 15], and the failure mechanism of the basic component that triggers the failure of the complete system.

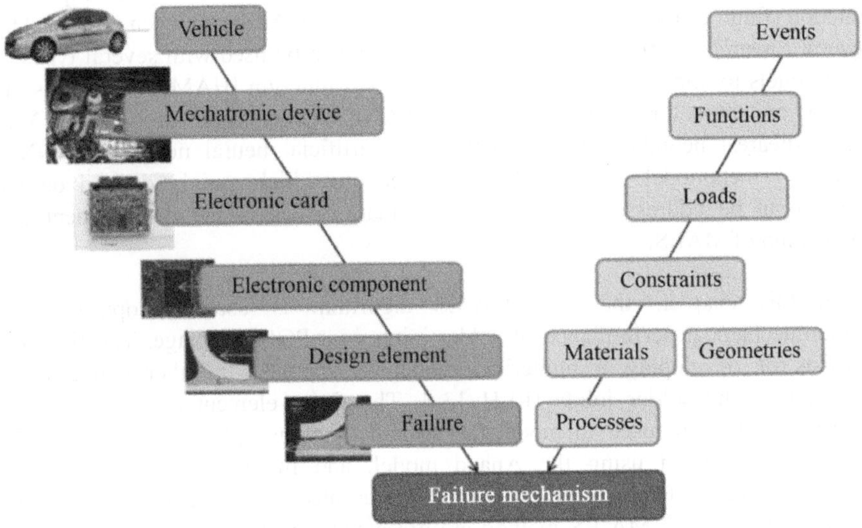

Figure 6.1. *Diagram of the components in the mechatronic system*

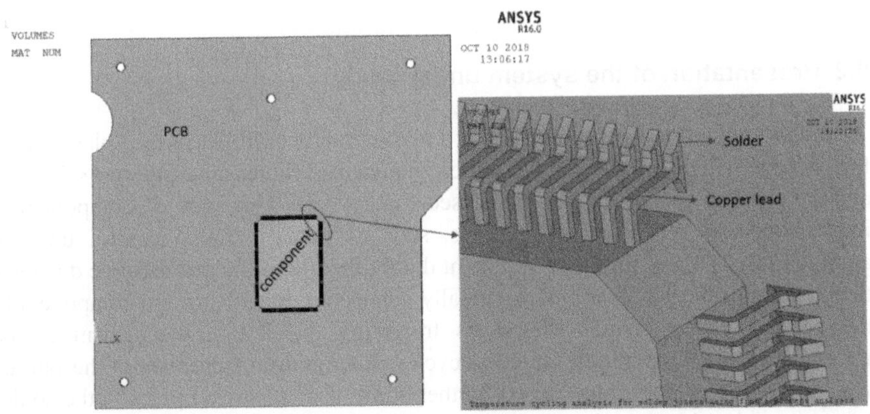

Figure 6.2. *PQFP housing structure (type of solder joints)*

For power mechatronic devices, the temperature range to which the devices are exposed is very wide. Therefore, the thermomechanical properties of the devices, especially those of the solder joints, are important for the reliability of the interconnected systems, as the devices are subjected to a very wide temperature variation [LIU 04]. Thermomechanical stress in solder joints due to the difference in material properties, especially the coefficient of thermal expansion (CTE) between pins, solder joints and substrate materials, can lead to damage and thus failure of the mechatronic device. The performance of these structures is also highly dependent on design parameters.

6.2.1. *The case of wire bonding*

Thermomechanical stress in wire bonds is due to the imbalance in the CTE between the wire, bonding pads and substrate materials. This difference can lead to microcracking at the solder joint between the bond wire and the chip and consequently cause the bond wire to become detached [LIU 04].

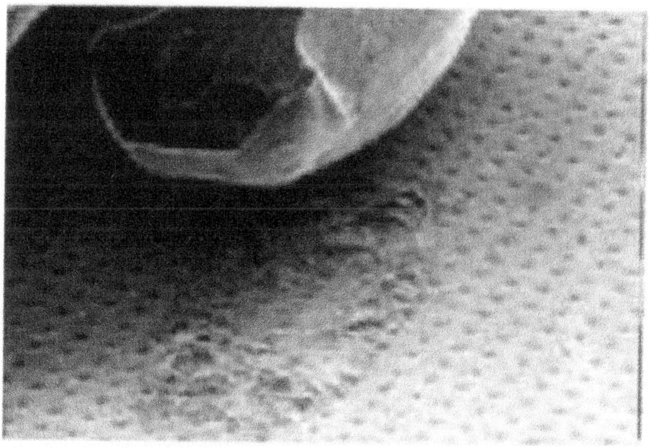

Figure 6.3. *Detachment of the wire bonding*

6.2.2. *The case of solder joints*

The solder joints used for mounting the packages on the printed circuit board (PCB) is a concrete example of an assembly with different CTEs (see Table 6.1). These joints are the means by which the package and the board are connected. However, with advances in miniaturization, in order to have smaller products, solder joint stresses become a major issue and challenge. This is the case for QFN, BGA, CSP, QFP, etc. The solder joints used can accept high strain and stress rates. These thermomechanical stresses result in the warping of the PCBs (see Figure 6.5). The solder joint stresses are mainly shear stresses, although tension/compression stresses are also present. However, a repetition of the solicitations (thermal cycles) leads to their ageing, and consequently, the rupture of the soldered joints (thermomechanical fatigue phenomenon) (see Figure 6.3).

Coefficient of thermal expansion (CTE) (in $ppm \cdot {}^{\circ}C^{-1}$)			
Alumina	7	Molybdenum	5.5
Aluminum	22	Gold	13.2
Beryllium	8	Porcelain	4.1
Copper	18	Polyimide	40–50
Diamond	1.18	Silicon	2.8
Kovar	6	Soft solder	~21
Epoxy glass	11–15 (XY plane) and 60–80 (Z axis)		

Table 6.1. *Coefficient of thermal expansion of materials used in microelectronics [POU 16]*

Figure 6.4. *Example of a solder joint failure*

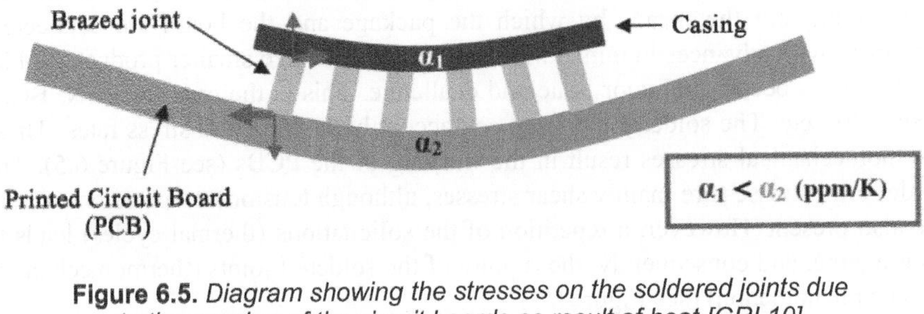

Figure 6.5. *Diagram showing the stresses on the soldered joints due to the warping of the circuit boards as result of heat [GRI 10]*

In order to achieve optimization and increase the performance of mechatronic structures, a large number of FE simulations will need to be run in the optimization algorithms. However, the use of FE simulations in the thermomechanical analysis of mechatronic devices is a complex nonlinear problem that takes time to execute. Therefore, the use of a FEM-based code in the optimization process will be costly in terms of computational time. The main objective is therefore to take advantage of the performance of metamodel-assisted optimization algorithms to develop a design optimization methodology to improve the performance and reliability of mechatronic components, while avoiding the computational cost generated by FEM. The performance improvement is achieved by extending fatigue life, which is calculated using thermal fatigue models.

6.3. Thermal fatigue models of solder joints

Fatigue models are used to predict the fatigue life of solder alloys. The models proposed in the literature can generally be classified into five main categories [LEE 00], according to the mechanism being considered, and accounting for the source of failure of the alloys: plastic deformation-based approach, stress-based approach, energy-based approach, creep-based approach and damage-based approach.

In the field of electronic reliability, power law models are the most widely used due to their simplicity of identification and application. The number of cycles to failure is expressed as a function of the level of cyclic loading applied. The empirical approach represents a research technique based on observation and experience, where the number of cycles to failure is expressed as a function of the level of cyclic stress applied. Their expression is given as follows:

$$\Delta X\left(N_f\right)^b = A \qquad [6.1]$$

where ΔX is the magnitude of the parameter indicating the level of stress, A is the fatigue ductility coefficient, N_f is the number of cycles to failure and b is the fatigue ductility exponent.

The determination for the number of N_f fatigue lifecycles requires the choice and knowledge of a physical cyclic parameter (ΔX). The physical parameters employed (see Figure 6.6) are as follows:

– the amplitude of inelastic deformation ($\Delta\gamma_{in}$) [MAK 16];

– the amplitude of total deformation ($\Delta\gamma_t$) [COF 54];

– the plastic work density per cycle (W_p) [DAR 00];

– the cumulative viscoplastic strain per cycle [SEY 04] (calculated via behavior laws).

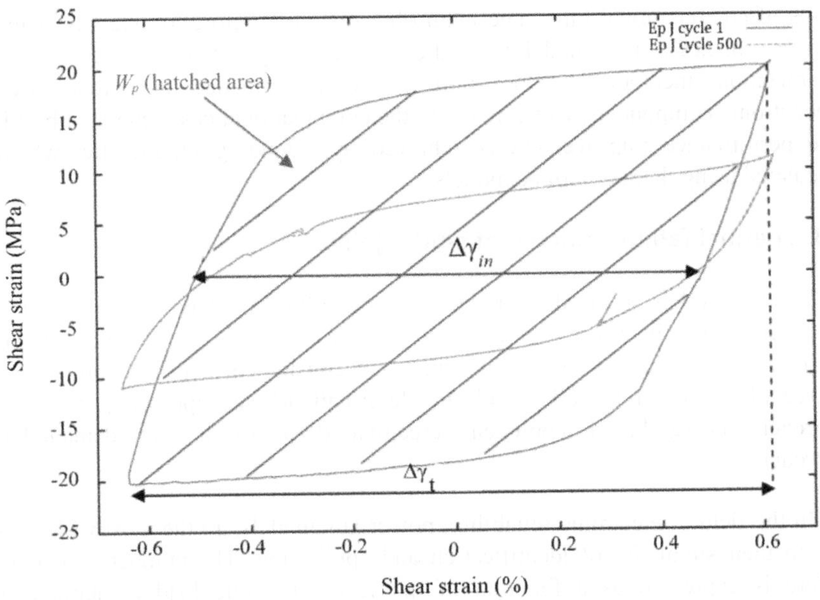

Figure 6.6. *Stress–strain evolution of a shear torsion specimen of a SAC305 alloy at 25°C in low-cycle fatigue for a frequency of 0.05 Hz for $\Delta\gamma_t = 1.3\%$ of the first and 500th cycles*

The following sections present the main types of fatigue models.

6.3.1. *The Coffin–Manson model*

This is one of the most widely used models for fatigue life prediction [COF 54]. The total number of cycles to failure is expressed as a function of the plastic strain amplitude. It is usually expressed by the following equation:

$$\Delta\varepsilon_p N_f^{\alpha} = \theta \qquad [6.2]$$

where $\Delta\varepsilon_p$ is the magnitude of the parameter indicating the level of stress, N_f is the number of cycles to failure, θ is the fatigue ductility coefficient and α is the fatigue ductility exponent.

The fatigue ductility coefficient θ and the fatigue ductility exponent α are determined experimentally. Table 6.2 shows example values of these two parameters for three types of solder alloys.

Alloy	α	θ
SnPb	0.68	0.85
SAC305	0.73	3.7
SnAg3.5	0.93	21.9

Table 6.2. *Parameters for the Coffin–Manson model*

The amplitude of the plastic shear strain $\Delta\gamma_p$, or amplitude of the total strain $\Delta\gamma_t$, can be used instead of the amplitude of the plastic strain $\Delta\varepsilon_p$ [SOL 89]. In the case of multiaxial deformations, it is recommended to adopt an equivalent deformation expressed as follows [BAN 90]:

$$\Delta\varepsilon_p = \frac{\sqrt{2}}{3}\left[\left(\Delta\varepsilon_1 - \Delta\varepsilon_2\right)^2 + \left(\Delta\varepsilon_2 - \Delta\varepsilon_3\right)^2 + \left(\Delta\varepsilon_3 - \Delta\varepsilon_1\right)^2\right] \quad [6.3]$$

6.3.2. *The Morrow model*

The Morrow's [MOR 65] model uses the inelastic work dissipated during a cycle W_{in} as a physical cyclic parameter instead of the plastic strain. This model has the advantage of taking into account both strain and stress. It is expressed in the following form:

$$W_{in} \cdot N_f^{\alpha} = \theta \quad [6.4]$$

where W_{in} is the inelastic work dissipated during one cycle (MPa), θ is the fatigue ductility coefficient, N_f is the number of cycles to failure and α is the fatigue ductility exponent.

6.3.3. *The Coffin–Manson frequency-modified model*

In this model, the frequency of solicitation is integrated into the model and $N_f v^{k-1}$ is defined as the "modified frequency lifetime". This variable can be expressed differently depending on the frequency range as proposed by Pang et al. [PAN 04], by introducing two parameters k_1 and k_2. The exponent k depends on the temperature [SHI 00] for tests at constant temperature. Taking into account the frequency allows us to take into account the strain rate, which considerably affects the behavior and the damage of highly viscoplastic alloys.

$$\left(N_f v^{k-1}\right)^{\alpha} \Delta\varepsilon_p = \theta \qquad [6.5]$$

where $\Delta\varepsilon_p$ is the amplitude of plastic deformation, v is the frequency of the stress (Hz), k is the frequency exponent, θ is the fatigue ductility coefficient (unit depends on k), N_f is the number of cycles until failure and α is the fatigue ductility exponent.

6.3.4. *The Morrow frequency-modified model*

Similar to the Coffin–Manson frequency-modified model, frequency is similarly incorporated into this model and $N_f v^{k-1}$ is referred to as the frequency-modified lifetime. The model is expressed in the following form:

$$\left(N_f \mathbf{y}^{k-1}\right)^{\alpha} W_{in} = \theta \qquad [6.6]$$

where W_{in} is the inelastic strain energy density accumulated during one cycle (J.m^{-3}), v is the frequency of the load (Hz), k is the frequency exponent, θ is the fatigue ductility coefficient (unit depends on k), N_f is the number of cycles until failure and α is the fatigue ductility exponent.

6.3.5. *The Darveaux model*

Makhloufi et al. [MAK 16] were able to establish two equations as well as four crack growth correlation constants K_1 to K_4, where the results of the finite element simulations can be used to calculate the crack propagation rate per thermal cycle as well as the number of cracking cycles. Equations [6.7] and [6.8] can be used to calculate the number of cycles for macroscopic crack initiation "N_0" and the evolution of the crack length as a function of the number of cycles "da / dN", respectively. Darveaux shows that the values of the constants depend on the thickness of the computational zone selected in the finite element models.

The characteristic fatigue life of the solder joint "α" (number of cycles to population failure) is expressed as the sum of "N_0" and the number of cycles for crack propagation over the entire diameter of the solder joint "a", as shown in equation [6.9].

$$N_0 = K_1 \left(W_{ave}\right)^{K_2} \quad [6.7]$$

$$\frac{da}{dN} = K_3 \left(W_{ave}\right)^{K_4} \quad [6.8]$$

$$\alpha = N_0 + \frac{a}{da/dN} \quad [6.9]$$

where N_0 is the number of cycles before crack initiation, W_{ave} is the average plastic work density per cycle (psi), a is the frequency exponent, K_1 is the material constant (cycles/psi^{K2}), K_2 is the material constant, K_3 is the material constant ($10^{-7}\,\text{in/cycle/psi}^{K4}$) and K_4 is the material constant.

6.4. Modeling and finite element analysis of the PQFP housing

6.4.1. *Modeling*

Finite element analysis is used to quantify the lifetime of the PQFP component. However, the calculation using the global model (see Figure 6.7) of the circuit board with the microcontroller [ELH 15] has a high calculation cost. To this end, a local 3D model has been developed to better account for the most critical joint, so as to reduce the overall computation time (see Figure 6.8).

The local model of the critical area is developed, which is the solder joint most stressed by the thermal loading. This local sub-model (see Figure 6.8) is composed of a FR4 printed circuit board (PCB), an EPOXY resin component, a copper pin and a SAC305 type solder joint. Symmetry boundary conditions are introduced in the modeling and the Anand model is used to account for the viscoplastic behavior of the solder joint [GRI 10]. All models in this study are developed with the $ANSYS^{TM}$ finite element method simulation tool.

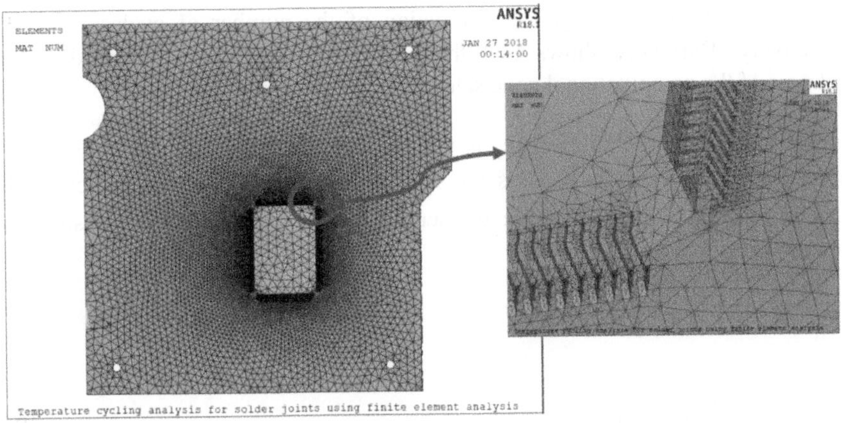

Figure 6.7. *Global model of solder joints and the adopted mesh size*

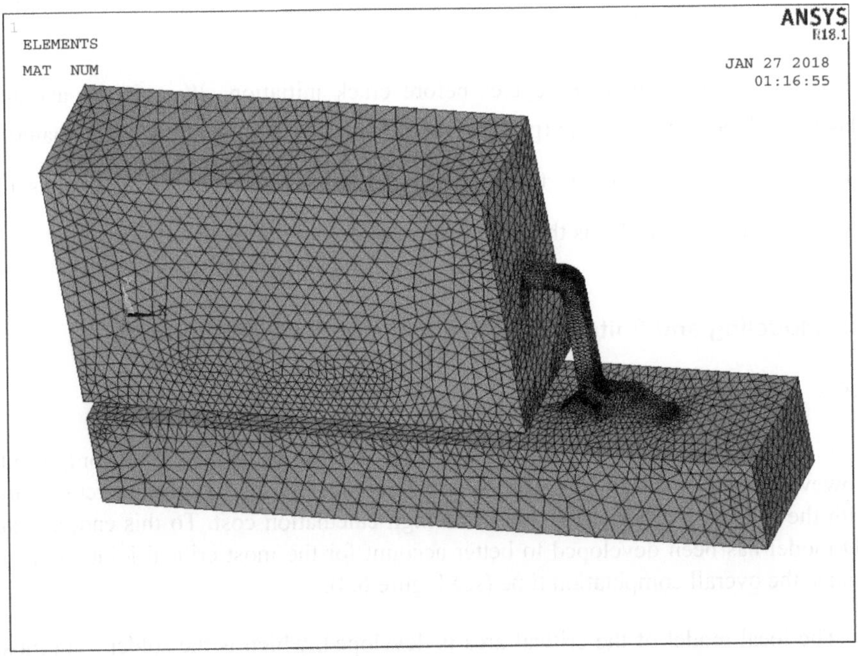

Figure 6.8. *Local model of solder joints*

6.4.2. *Material properties*

In the thermomechanical analysis using the local model, the solder joint material is assumed to have viscoplastic behavior. Several authors have studied the response of lead-free solder joints (SnAgCu) and proposed equations to model this response. One of the models developed is the Anand model, which distinguishes between elastic and inelastic deformations, and combines creep and instantaneous plastic deformations [WAN 01]. Moreover, the Anand model also expresses the viscoplastic behavior of the material and its equation is defined as follows:

$$\dot{\varepsilon}_p = A\exp\left(\frac{-Q}{RT}\right)\left[\sinh\left(\xi\frac{\sigma}{s}\right)\right]^{\frac{1}{\mathrm{m}}} \qquad [6.10]$$

where A is a pre-exponential factor (1/s), T is the temperature, R is the universal gas constant, ξ is the material constant and m is the strain sensitivity of the stress.

The internal variable s represents the resistance to plastic deformation, and its equation evolution defined by:

$$\dot{s} = \left\{h_0\left|1-\frac{s}{s^*}\right|^a .\operatorname{sign}\left(1-\frac{s}{s^*}\right)\right\}.\dot{\varepsilon}_p \quad ; \quad a>1 \qquad [6.11]$$

with:

$$\hat{s}\left(\frac{\dot{\varepsilon}_p}{A}\exp\left(\frac{Q}{RT}\right)\right)^n \qquad [6.12]$$

The nine parameters A, Q/R, $\hat{s}$, h_0, ξ, m, n, a and s_0, which are required to identify the evolution of the strain resistance in equation [6.11] and the strain rate in equation 6.12, are shown in Table 6.3.

Table 6.4 shows the material properties of the local model used in this study.

Parameter		Value	Description
A	(s^{-1})	2.2310^{4}	Pre-exponential factor
Q / R	(K)	8900	Q = Activation energy/ R= Boltzmann constant
ξ	(without)	6	Stress multiplier
m	(without)	0.182	Stress sensitivity rate
$\hat{s}$	(without)	73.81	Coefficient of the saturation value to the deformation strength.
n	(without)	0.018	Strain rate sensitivity
h_0	(MPa)	3321.15	Hardening constant
a	(without)	1.82	Rate of sensitivity to the strain of hardening.
s_0	(MPa)	39.09	Initial value of the resistance to deformation

Table 6.3. *Parameters of the Anand model [GRI 10]*

Parameter		SAC305	FR4	Epoxy Resin	Cu
Young's modulus	GPa	51.3	17	17	115
Poisson coefficient	dimensionless	0.3	0.3	0.2	0.31
Density	kg/m^3	740	180	180	8.890
CTE	µm/K	20	18	22	17
Shear modulus	GPa	19	2.4	7.4	44

Table 6.4. *Material properties*

6.4.3. *Thermal load*

Accelerated thermal cycling, recommended by JEDEC standards [JES 00], is applied as thermal loading in FEM analysis. The temperature profile applied ranges from –40°C to 125°C, for 30 minutes at low and high residence times. Devices for automotive applications are typically tested in this temperature range. Figure 6.9 shows the thermal loading cycle. In the FEM calculation, the first step is to simulate the reflow soldering process to account for the initial stresses. This process involves applying a temperature profile from the solder joint melting temperature to the ambient temperature of 25°C within 150 s. The second step is to apply three thermal cycles, between –40°C and 125°C, as the loading condition for the FEM simulation.

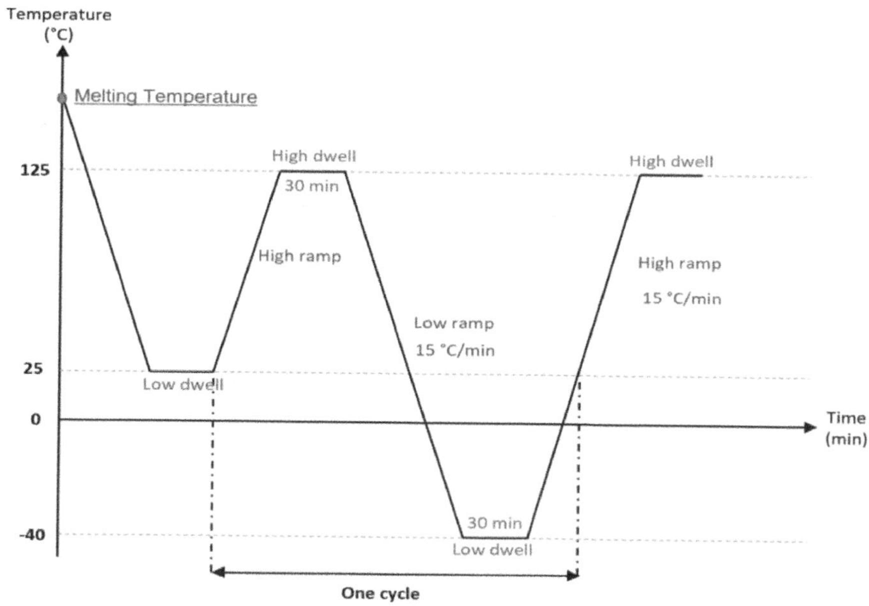

Figure 6.9. *Thermal load cycle*

6.4.4. *Fatigue model selected for solder joints*

Prediction of solder joint fatigue requires a combination of the finite element method with a thermal fatigue model, which is usually obtained from experimental data and accelerated testing. This model is used to calculate the number of cycles the system can withstand before failure. Of the models listed above, the Coffin–Manson fatigue model, classified under the plastic deformation approach, is the best and most commonly used approach. The number of cycles N_f is expressed as a function of the inelastic deformation range $\Delta\varepsilon_{in}$, the fatigue ductility exponent $\alpha\,(0.73)$ and the fatigue ductility coefficient θ $(\theta = 3.7)$ [KAN 02]. The form of the Anand model used in this analysis is given by the following equation:

$$\Delta\varepsilon_{in} N_f^{\alpha} = \theta \qquad [6.13]$$

The inelastic deformation range $\Delta\varepsilon_{in}$ is calculated using a nonlinear finite element simulation based on the Anand model.

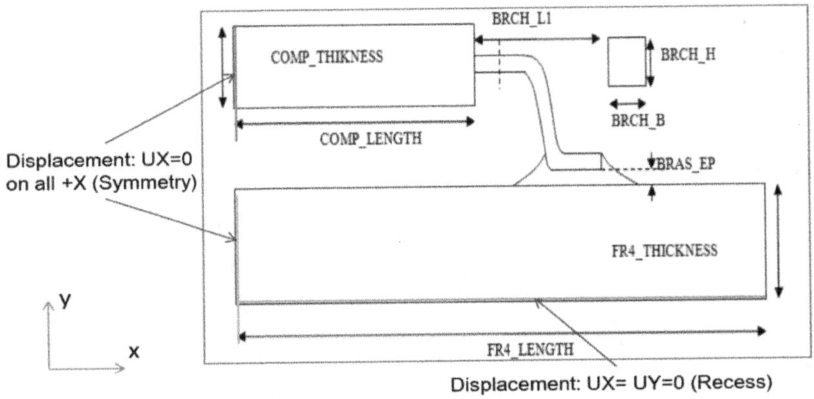

Figure 6.10. *Boundary conditions applied to the sub-model*

6.4.5. *Numerical results*

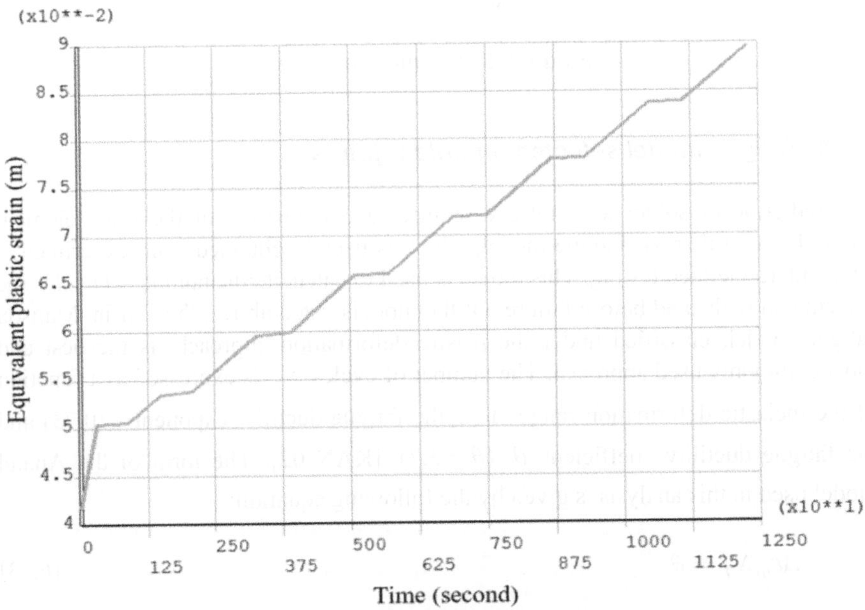

Figure 6.11. *Plastic deformation history*

Once the sub-model design is completed as shown in Figure 6.8, and the boundary constraints have been applied as shown in Figure 6.10, the solution is configured by the ANSYS™ commands. Finally, three thermal cycles are simulated. Each thermal cycle consists of four load levels (high ramp, high stay, low ramp and low stay). Therefore, a complete simulation of three thermal cycles contains 12 loading stages. The three thermal cycles are applied to achieve the stabilization of the inelastic deformation generated by each thermal cycle, as shown in Figure 6.11.

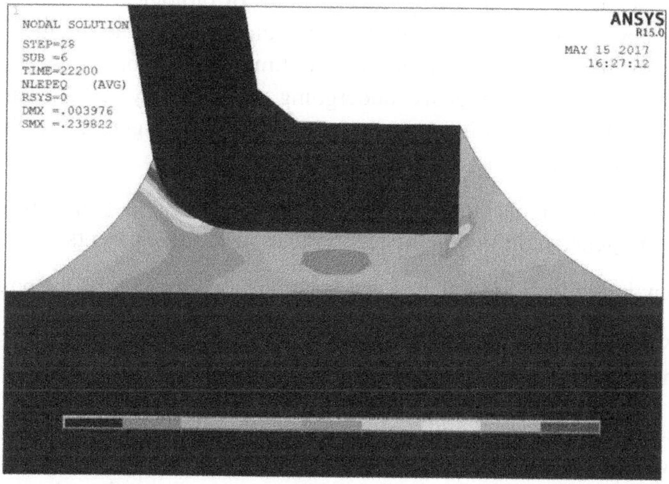

Figure 6.12. *Side view of the plastic strain distribution in the solder joint after the final loading cycle*

6.5. Evolutionary strategies

6.5.1. *Presentation of evolutionary strategies*

ESs are derivative-free stochastic optimization methods. These methods are appropriate and effective for continuous black-box optimization, that is, for optimization scenarios where no analytical expression for the objective function is explicitly given and derivatives are not available [MAK 16]. Furthermore, ESs have the ability to converge to a global optimum for multimodal problems, making ESs increasingly popular for solving real-world optimization problems.

ESs belong to the family of evolutionary algorithms (EAs). They are search models inspired by the principles of biological evolution. Their principle is to approach optimization problems by implementing a repeated process of stochastic variations followed by a selection. At each iteration, new individuals, called children (solutions) are generated from their parents (solutions already obtained), next, their image by the objective function is evaluated and the best individuals are selected to become the parents for the next generation [HAN 15].

6.5.2. *Principles of ESs*

ESs are mechanisms inspired by the principles of biological evolution that use a repeated process composed of evolutionary operators that mimic the mechanisms of the Darwinian theory of evolution to solve optimization problems. A population P consists of a set of individuals undergoing evolutionary cycles during each generation (iteration) of evolutionary search. In general, a generation loop can be summarized in three steps:

– one or more parents are selected from the population (selection) and new individuals are generated through the recombination of these parents;

– the new individuals are mutated and become new members of the population;

– a selection reduces the population to its original size by choosing the best individuals.

This sequence is repeated until one of the convergence criteria of the algorithm is satisfied, which can be reaching a maximum number of evaluations of the objective function, reaching a target value of the function, or stagnating the evolutionary process.

In order to implement EAs, the principles described above must be specified as concrete evolutionary operators to form evolutionary search patterns. The operators of selection, recombination and mutation are the three operators used.

6.5.3. *Covariance matrix adaptation evolution strategy*

The covariance matrix adaption evolution strategy (CMA-ES) method is one of the best ESs for derivative-free optimization. It has become a standard for black-box continuous evolutionary optimization. The correlated mutation operator adopted by CMA-ES makes it a powerful optimization algorithm when compared to other algorithms that use isotropic mutations. CMA-ES is based on a multivariate normal distribution parameterized as a distribution model for a population of candidates, at each generation of CMA-ES, the descendant candidates λ are generated from the parents μ. In the next generation, the new set of parents (μ) are selected from the descendant candidates (λ) by the selection of (λ, μ). The best candidates are selected based on their ranking according to the objective function. The CMA-ES algorithm (λ, μ) uses two techniques, an adaption of the covariance matrix adaptation (CMA) and an adaption of the cumulative step-size adaptation (CSA).

More specifically, let $\mathcal{N}(m^{(g)}, K^{(g)})$ be a multivariate normal distribution with a mean of $m^{(g)}$ and a covariance matrix of $K^{(g)}$. In CMA-ES, the mean sequence $m^{(g)}$ represents the preferred solution at generation g or the best estimate of the optimum. In generation g, the covariance matrix g is factorized into two parts: $C^{(g)} \in \mathbb{R}^{n \times n}$ and $\sigma^{(g)} \in \mathbb{R}^{+}$, where $C^{(g)}$ represents a covariance matrix (positive definite matrix) and $\sigma^{(g)}$ is a scalar factor to control the step length of the iteration.

6.5.3.1. *Mutation or sampling*

In the CMA-ES method (λ, μ), the λ individuals are generated according to the following equation:

$$x_k^{(g)} = m^{(g)} + \sigma^{(g)} \mathcal{N}_d(0, C^{(g)}) = m^{(g)} + \sigma^{(g)} \left(C^{(g)}\right)^{\frac{1}{2}} z_k \quad \text{for } k = 1, \ldots, \lambda \qquad [6.14]$$

where:

– $g \in \mathbb{N}^0$ is the counter for the number of generations;

– $\lambda \geq 2$ is the population size (number of individuals). Its default value is $\lambda = 4 + \lfloor 3ln(d) \rfloor$ (and where d is the dimension of the problem);

– x_k is the kth individual;

– $m^{(g)}$ is the mean vector of the distribution (solution), formed from a weighted recombination of the best individuals μ after initialization. Its initial value $m^{(0)}$ is given by the user;

– $\sigma^g \in \mathbb{R}^{+}$ is the step size at generation g. Its initial value is given according to the bounds of the search space;

– $C^g \in \mathbb{R}^{d \times d}$ is the covariance matrix (symmetric positive definite) at generation g. Its initial value is $C(0) = \left(C^{(g)}\right)^{\frac{1}{2}} \in \mathbb{R}^{d \times d}$, the square root matrix of $C^{(g)}$;

– $z_k \in \mathbb{R}^d$ are the random vectors sampled from $\mathcal{N}_d(0,1)$, that is, $z_k = \mathcal{N}_d(0,1)$.

To completely define the stochastic iterative algorithm of CMA-ES, the tree terms, denoted as $m^{(g)}$, $C^{(g)}$ and $\sigma^{(g)}$, require an update at each generation.

6.5.3.2. *Selection and recombination*

After the sampling process, the children λ are evaluated and ranked according to their images by the objective function. Finally, the best individuals μ are selected from λ and the average $m^{(g+1)}$ is then updated by taking the weighted average of μ individuals.

$$m^{(g+1)} = \sum_{i=1}^{\mu} w_i \mathbf{x}_{i:\lambda}^{(g)} = m^{(g)} + \sigma^{(g)} \left(C^{(g)}\right)^{\frac{1}{2}} z_{i:\lambda} \qquad [6.15]$$

$$\sum_{i=1}^{\mu} w_i = 1, \quad w_1 \geq w_2 \geq \cdots \geq w_\mu > 0 \qquad [6.16]$$

where:

– μ is the number of selected individuals with a default value equal to $\mu = \lfloor \lambda / 2 \rfloor$;

– $(w_i)_{1 \leq i \leq \mu}$ is the μ strictly positive recombination weights, satisfying $\sum_{i=1}^{\mu} (w_i)_{1 \leq i \leq \mu} = 1$. In the case where $w_i = 1/\mu, i = 1, \cdots, \mu$, it is the intermediate recombination, and for the standard CMA-ES, the superlinearly weights are used:

$$w_i = \left[\ln(\lambda/2 + 1/2) - \ln(i)\right] \Big/ \sum_{j=1}^{\mu} \left[\ln(\lambda/2 + 1/2) - \ln(j)\right], \; i = 1, \cdots, \mu; \qquad [6.17]$$

– $\mathbf{x}_{i:\lambda}$ is the ith best individual selected from the set of individuals $\mathbf{x}_1, \mathbf{x}_2, \cdots, \mathbf{x}_\lambda$ generated by equation 6.14. $\mathbf{x}_1, \mathbf{x}_2, \cdots, \mathbf{x}_\lambda$ $i:\lambda$ is the index of the ith individual after ranking, such that $f(x_{1:\lambda}) \leq f(x_{2:\lambda}) \leq \cdots \leq f(x_{\lambda:\lambda})$, in which f is the objective function;

– $z_{i:\lambda}$ is the mutation vector associated with $\mathbf{x}_{i:\lambda}$.

The CMA-ES method uses cumulative step-size adaptation (CSA) and covariance matrix adaptation (CMA) successively to update the step size σ and covariance matrix C. Both techniques use so-called evolution paths to accumulate information about the parameters of the strategy across generations. The two evolution paths used in CMA-ES are $p_\sigma \in \mathbb{R}^d$ (see equation 6.18) for the step size adaptation and $p_c \in \mathbb{R}^d$ (see equation 6.21) for the adaptation of the covariance matrix C.

The evolution path $\mathbf{p}_\sigma^{(g)} \in \mathbb{R}^d$ and step size σ are calculated according to the following two equations, respectively:

$$\begin{aligned}\mathbf{p}_\sigma^{(g+1)} &= (1-c_\sigma)\mathbf{p}_\sigma^{(g)} + \sqrt{c_\sigma(2-c_\sigma)}\sqrt{\mu_w}\left(\mathbf{C}^{(t)}\right)^{-\frac{1}{2}}\frac{\mathbf{m}^{(g+1)}-\mathbf{m}^{(g)}}{\sigma^{(t)}} \\ &= (1-c_\sigma)\mathbf{p}_\sigma^{(g)} + \sqrt{c_\sigma(2-c_\sigma)}\sqrt{\mu_w}\sum_{i=1}^{\mu} w_i \mathbf{z}_{i\lambda}\end{aligned} \qquad [6.18]$$

$$\sigma^{(g+1)} = \sigma^{(g)} \exp\left(\frac{c_\sigma}{d_\sigma}\left(\frac{\left\|\mathbf{p}_\sigma^{(g+1)}\right\|}{\mathbb{E}\left(\left\|\mathcal{N}_d(\mathbf{0},\mathbf{I})\right\|\right)} - 1\right)\right) \qquad [6.19]$$

where:

– $p_\sigma^{(g)} \in \mathbb{R}^d$ is the evolution path for the step size at generation g, and its initial value is $\mathbf{p}_\sigma^{(0)} = 0$;

– $c_\sigma \in [0,1]$ is the time constant for step size adaptation, and its default value is $c_\sigma = (\mu_w + 2)/(d + \mu_w + 5)$;

– c_σ is called variance effective selection mass, and its value is $\mu_w = \left(\sum_{i=1}^{\mu} w_i\right)^2 / \sum_{i=1}^{\mu} w_i^2$;

– $\left(\mathbf{C}^{(t)}\right)^{-\frac{1}{2}}$ is the inverse of $\left(\mathbf{C}^{(t)}\right)^{\frac{1}{2}}$, the symmetrical square root matrix of $C^{(g)}$;

– d_σ is the damping factor, with a default value defined as follows: $d_\sigma = 1 + 2\max\left(0, \sqrt{\frac{\mu_w - 1}{d+1}} - 1\right) + c_\sigma$;

– $\mathbb{E}\left(\left\|\mathcal{N}_d(\mathbf{0},\mathbf{I})\right\|\right)$ is the expectation of the Euclidean norm for the random vector $\mathcal{N}_d(\mathbf{0},\mathbf{I})$, whose approximation is given as follows:

$$\mathbb{E}\left(\left\|\mathcal{N}_d(\mathbf{0},\mathbf{I})\right\|\right) = \sqrt{2}\Gamma\left(\frac{d+1}{2}\right) / \Gamma\left(\frac{d}{2}\right) \approx \sqrt{d}\left(1-\frac{1}{4d}+\frac{1}{21d^2}\right). \qquad [6.20]$$

The evolution path $\mathbf{p}_c$ and the matrix C are updated, respectively, according to the following equations:

$$\begin{aligned}\mathbf{p}_c^{(g+1)} &= \left(1-c_c\right)\mathbf{p}_c^{(g)} + h_\sigma\sqrt{c_c\left(2-c_c\right)}\sqrt{\mu_w}\,\frac{\mathbf{m}^{(g+1)}-\mathbf{m}^{(g)}}{\sigma^{(g)}} \\ &= \left(1-c_c\right)\mathbf{p}_c^{(g)} + h_\sigma\sqrt{c_c\left(2-c_c\right)}\sqrt{\mu_w}\sum_{i=1}^{\mu} w_i\left(\mathbf{C}^{(g)}\right)^{\frac{1}{2}}\mathbf{z}_{i:\lambda}\end{aligned} \qquad [6.21]$$

$$\begin{aligned}C^{(g+1)} &= \left(1-c_1-c_u\right)\mathbf{C}^{(g)} + c_1\left(\mathbf{p}_c^{(g+1)}\left(\mathbf{p}_c^{(g+1)}\right)^{\mathrm{T}} + \delta\left(h_\sigma\right)\mathbf{C}^{(g)}\right) \\ &\quad + c_\mu\sum_{i=1}^{u} w_i\left(\left(\mathbf{C}^{(g)}\right)^{\frac{1}{2}}\mathbf{z}_{i,\lambda}\right)\left(\left(\mathbf{C}^{(g)}\right)^{\frac{1}{2}}\mathbf{z}_{i:\lambda}\right)^{T}\end{aligned} \qquad [6.22]$$

where:

– $p_c^{(g)} \in \mathbb{R}^d$ is the evolution path for the step size at generation g, and its initial value is $\mathbf{p}_c^{(0)} = 0$;

– $c_c \in [0,1]$ is the adaptation constant of the covariance matrix, and its default value is $c_c = \left(4+\mu_w / d\right) / \left(d+4+2\mu_w / d\right)$;

– h_σ is the Heaviside function defined by the following equation:

$$h_\sigma = \begin{cases} 1 & \text{if } \dfrac{\left\|\mathbf{p}_\sigma^{(g+1)}\right\|.}{\sqrt{1-\left(1-c_\sigma\right)^{2(g+1)}}} < \left(1.4+\dfrac{2}{d+1}\right)\mathbb{E}\left(\left\|\mathcal{N}_d(0,1)\right\|\right); \\ 0 & \text{if not} \end{cases} \qquad [6.23]$$

– c_1 and c_μ express the learning rate for rank-one-update and rank-μ-update $\left(\mathbf{C}^{(t)}\right)^{\frac{1}{2}}$, respectively, their default values are:

$$c_1 = \frac{2}{(d+1.3)^2 + \mu_w}, \text{ and } c_u = \min\left(1 - c_1, 2\frac{\mu_w - 2 + 1/\mu_w}{(d+2)^2 + \alpha_u \mu_w / 2}\right) \text{ with } \alpha_\mu = 2.$$

$$\delta(h_\sigma) = (1 - h_\sigma) c_c (2 - c_c).$$

At each iteration of CMA-ES, the new population of individuals is generated by equation [6.14]; the selection and recombination are performed by equation [6.15]; the adaptation of the step size is performed by equations [6.18] and [6.19]; and the adaptation of the covariance matrix is performed by equations [6.21] and [6.22]. The iteration defined previously is repeated until one of the end criteria is satisfied.

6.5.4. *Metamodeling techniques*

Metamodeling techniques are approaches to model the behavior (input–output) of a simulation model. Computer simulation tools, such as finite element (FE) analysis and computational fluid dynamics (CFD), are widely used in the analysis of engineering problems. The majority of engineering design problems require a number of simulations to achieve the design goals. However, performing a single simulation can take several minutes, hours or even days. As a result, computationally intensive tasks such as design optimization, sensitivity analysis and reliability analysis become impractical or impossible because they require hundreds, thousands or even millions of simulations. This tedious computation can be reduced by building approximation models, known as surrogate models or metamodels, which approximate the input–output relationship of the simulation model while being computationally inexpensive to evaluate [FAN 05]. Many tasks can be performed using the constructed metamodel with the bonus that the computational cost becomes far less expensive.

The general challenges for building a metamodel can be solved by using an approach that calls upon several techniques from various scientific fields. This approach can be summarized in the following steps:

– Experiment design: this is a sampling procedure that allows us to obtain a set of points $\mathbf{x}^{(i)}$ of the design space $\mathcal{D}$. Experiment designs are generated by several methods.

– Correct solution evaluation: this is the procedure that finalizes the second part of the learning base E_a; this phase uses the mechanical model to compute the responses of the objective function to the points obtained either through the iterative process (experimental designs) or through the enrichment procedure.

– Metamodel construction: this is the main phase in the development process. It requires the use of methodologies specific to the metamodel being employed. Most of the time, during the construction phase, a certain number of operators are constructed and these remain constant during the evaluation of the metamodel (i.e. to obtain $\tilde{y}\left(\mathbf{x}^{(0)}\right)$, where $\mathbf{x}^{(0)}$ is any point of $\mathcal{D}$). The construction phase can be particularly long depending on the metamodel.

– Enrichment: this is a procedure that allows new sites to be added to the learning base. This addition can be done on the existing design (lack of sites in a zone of the design space) or by using the information provided by the metamodel in a relevant way (by using an enrichment indicator based on the response of the metamodel, or even an error indicator provided by the latter).

– Validation: this is a convergence test of the metamodel. The validation enables measuring the difference between the solutions of the metamodel and those of the exact model in some points. This phase allows us to decide if the metamodel is good or not, and in the case where the considered approach integrates an enrichment procedure, this test can be used as a decision criterion on the continuation, or not, for additional new points to construct the metamodel.

Once the metamodel is constructed, it is then usable and can provide approximate solutions at any point in the design space, at an extremely low cost.

6.5.4.1. *The Kriging metamodel*

Kriging is a geostatistical technique for interpolating deterministic data without noise. The theoretical approach of Kriging comes from the work of Krige [KRI 51] and was formalized by the mathematician [MAT 63]. Subsequently, Kriging has become a reference method for the construction of metamodels for the numerical design of experiments [HAM 17]. It is used to predict the value of a costly function (FE model). This method is also known as the Gaussian process.

Let $y(x)$ be a function of x, with $x \in R_d$ and y a vector containing the n observations of $y(x)$ for the experimental design $D = \{x_1,\dots,x_n\}$, of dimensions $n * d$. The Kriging technique assumes that the function $y(x)$ is the realization of a Gaussian process, denoted as $Y(x)$, which is given by:

$$Y(x) = h(x) + Z(x) \tag{6.24}$$

where $h(x)$ is the mean of the process, and $Z(x)$ is a Gaussian process with zero mean and covariance, expressed as:

$$Cov\left(Z\left(x^{(i)}\right),Z\left(x^{(j)}\right)\right)=\sigma^2 R\left(x^{(i)},x^{(j)}\right) \text{ for } i,j=1,..,N \qquad [6.25]$$

where σ^2 is the variance of the Gaussian process, and $R\left(x^{(i)},x^{(j)}\right)$ its correlation function between any two samples $x^{(i)}$ and $x^{(j)}$.

The choice of the correlation function is an important element in Kriging construction. Many covariance functions are proposed in the literature [RAS 06] such as Matter, Gaussian, exponential or spherical correlation functions. The Gaussian correlation function is the most commonly used one; the latter enables us to control both the extent of influence and the regularity of the approximation model:

$$R(x^{(i)},x^{(j)})=\exp\left[\sum_{k=1}^{d}-\theta_k\left|x_k^{(i)}-x_k^{(j)}\right|^2\right] \qquad [6.26]$$

where d is the dimension of the design space, θ_k $(k=1,2,...,d)$ are unknown parameters of the correlation function, and where $x_k^{(i)}$ and $x_k^{(j)}$ are, respectively, the kth component for the sample point x_i and x_j.

For any new point x, the mean and variance of the prediction [SAC 89] can be calculated by:

$$\hat{y}(x)=h^T(x)\beta+r^T(x)R^{-1}(y-H\beta) \qquad [6.27]$$

$$s^2(x)=\sigma^2\left[r^T(x)Rr(x)+\frac{1-I^T(x)R^{-1}r(x)}{1-I^T(x)R^{-1}I}\right] \qquad [6.28]$$

where $\hat{y}(x)$ and $s^2(x)$ are, respectively, the estimated mean value and variance of $\hat{y}(x)$, $h^T=[h_i]$, $i=1,\ldots,k$, is a set of basis functions (e.g. polynomial functions), $\beta=(\beta_1,...,\beta_k)$ are the associated regression coefficients, R is the correlation matrix of D, H the matrix corresponding to the values of $h^T(D)$ and $r(x)=[R(x,x_1),\ldots,R(x,x_k)]^T$ is the vector of correlation functions between the untried point x and k samples $x^{(1)},...,x^{(k)}$. The unknown parameters of the metamodel can be determined using the likelihood estimation method (LEM) [ELH 11].

6.5.5. *Kriging-assisted CMA-ES*

In summary, proper management of the metamodel, when used with care to maintain high-quality throughout the optimization process, is the main strength of using metamodels with evolutionary strategy algorithms. In this section, metamodel management and quality will be discussed and elements of the KA-CMA-ES will be presented [HUA 18].

6.5.5.1. *Management and measuring quality of Kriging*

6.5.5.1.1. Management of Kriging

In evolutionary computation, the use of the metamodel to approximate the objective function is not as straightforward as we might suppose. Its use raises two major concerns.

– the convergence of the EA to the global optimum of the objective function should be ensured;

– the computational cost of the EA, which is the main goal behind the application of the metamodel, should be reduced as much as possible [JIN 05].

One of the main problems in replacing the objective function with a global metamodel is that it is very difficult to construct an accurate metamodel that can efficiently approximate the original function; it may happen that the EA converges to a false optimum [JIN 11]. To avoid this problem, the metamodel and the original function must be used together (in parallel); this means that for the computation, the original objective function is used to evaluate all individuals of a generation (generation control) or some of the individuals (individual control) [JIN 05]. This technique is the main issue in metamodel management, which is generally divided into three main approaches:

– the first is the so-called "uncontrolled evolution", where the metamodel is assumed to be of high quality and no individuals or generations are controlled, that is, the original objective function is not used in the evolutionary computation;

– the second is "fixed evolution control", where the frequency of evolution control is fixed. For this, there are generally two approaches, one is individual-based control, and the other is generation-based control [JIN 11];

– the last approach is the so-called "adaptive evolution control" where the evolution control depends on the quality of the metamodel.

6.5.5.1.2. Measuring the quality of the Kriging

In addition to reducing the computational cost, the quality or fidelity of the metamodel is the main issue in the construction and implementation of metamodels for evolutionary computation. Therefore, in order to measure the quality of the metamodel that avoids the problem of computational cost, it is not recommended to use a narrow quantitative approximation to evaluate its quality. In evolutionary computation, to be realistic, the metamodel should instead allow the EA to select the best individuals in terms of their images by the original objective function. In [JIN 03], several quality measures were introduced.

In CMA-ES, the objective function is evaluated for each individual in the generation and the progress of the algorithm is based mainly on the ranking of the individuals (selection by rank), which is the only information obtained from the evaluation of the objective function. In this sense, it is normal that the quality of the metamodel is measured by its ability to classify the individuals in the population [RUN 04].

The *approximate ranking procedure* [RUN 04], which can be classified as an adaptive evolutionary control mechanism, is one of the best methods to control the quality of the metamodel, through its ability to rank individuals in the population without using the exact ranking of the whole population.

In the approximate ranking procedure, a single individual is selected to be evaluated by the original function in each iteration, next the metamodel is updated and the site of the individuals is chosen based on the updated metamodel. This loop is stopped in the case where all individuals of the generation have been reevaluated, or if the site of the parent individuals does not change. Nevertheless, in the case of high-dimensional or multimodal problems, the population size is important. In addition, the amount of information added at each iteration may result in unimportant changes, even in the case of a metamodel with poor ranking predictions. To address this shortcoming, Huang et al. [HUA 18] propose some modifications of the original approximate ranking procedure:

– the first change is to select $n_{init} = \max(1, \lfloor 0.3\lambda \rfloor)$ individuals, instead of one, to be re-evaluated by the original function in order to update the metamodel. Notably, more information will be added by this change;

– the second modification is that, in the iteration loop of the approximate ranking procedure, the batch size $n_b = \max(1, \lfloor \lambda / 10 \rfloor)$ proportional to λ is used.

The main objective of the approximate ranking procedure is to determine the number of individuals to control in each generation. However, individuals are selected to be evaluated by the objective function based on their approximate rank (ranked by the metamodel) and then added to the training set $\mathcal{T}$. This operation is repeated until the selection of parent individuals by the metamodel remains unchanged for two successive iterations. In Algorithm 1, n_b is a batch of individuals to be evaluated for each iteration of the approximate ranking procedure, and the batch size used is $n_b = \lfloor \lambda / 10 \rfloor$ [KER 06].

Algorithm 1: Approximate ranking procedure

Approximate: construction $\hat{f}(x)$ based on the training site $\mathcal{T}$ and predict $\hat{f}(x_k), \quad k = 1, \ldots, \lambda$

Classify and determine the site of the parent individuals $\mathcal{P}_1 \equiv \left\{ x_{i:\lambda}^{(g+1)} \right\}_{i=1}^{\mu}$

Evaluate f (original function) by n_b best individuals (evaluation by the original function) and add the set to $\mathcal{T}$

For $n_{ic} = 2 \quad to \quad (\lambda / n_b)$ **perform**

    Approximate: construction $\hat{f}(x)$ based on $\mathcal{T}$ and predict $\hat{f}(x), \quad k = 1, \ldots, \lambda$

    Classify and determine the site of the parent individuals $\mathcal{P}_{n_{ic}} \equiv \left\{ X_{i:\lambda}^{(g+1)} \right\}_{i=1}^{\mu}$

    If $\mathcal{P}_{n_{ic}-1} \neq \mathcal{P}_{n_{ic}}$ **then**

        (the site of the parent individuals has changed)

        evaluate f of n_b best non-evaluated individuals and add to $\mathcal{T}$

    Otherwise (the site of the parent individuals remains unchanged)

        (Exit the loop)

    End if

End of

6.5.5.2. *KA-CMA-ES algorithm*

The chosen Kriging uses a Gaussian correlation function and quadratic basis functions. Therefore, in order to construct a Kriging metamodel, there are n_k required points, which represents the number of free parameters in the quadratic basis function $n_k = p = 1/2(n+1)(n+2)$. Therefore, the approximate ranking procedure will be performed in the case where the required number of points is assured, otherwise, the number of points in the training set $|\mathcal{T}|$ is greater than or equal to n_k.

The training data site $\mathcal{T}$ contains all previously evaluated individuals. However, it is not necessary to use all previously evaluated points as the training site, since the computational time to build a metamodel increases with the number of training points $|\mathcal{T}|$ [BRA 05]. In the Kriging-based CMA-ES algorithm, the main objective of the constructed metamodel is to predict the values, by the objective function, for the individuals (population) of the current generation; as such, it is thus reasonable to choose the current population as the training site. For the CMA-ES algorithm, where the multi-normal distribution is used, the Mahalanobis distance, $\Delta_{x_i}^2 = \left(x_i - m^{(g)}\right)^T \left(\left(\sigma^{(g)}\right)^2 C^{(g)}\right)^{-1} \left(x_i - m^{(g)}\right)$, is suitable for selecting the training data set [REN 03]. The Mahalanobis expression gives the distance between each evaluated point and the current mean $m^{(g)}$. Finally, the training set can be determined as follows:

$$\mathcal{T} \leftarrow \left\{(x, f) \in \mathcal{S} \mid \Delta_x^2 \leq \chi_n^2(p)\right\} \quad [6.29]$$

where $\mathcal{S}$ is an archive containing all previously evaluated points, and $\chi_n^2(p)$ is the quantile function for the probability p of the chi-square distribution with n degrees of freedom. It could be that $|\mathcal{T}| \leq n_k$ even if the number of training data $|\mathcal{S}|$ is greater than or equal to n_k. In this case, the required points n_k are selected from the $\mathcal{S}$ archive by modifying the quantile function $\chi_n^2(p)$ by r_{n_k}, the n_k points of $\mathcal{S}$ which have the smallest mahalanobis distance. Therefore, the training set $\mathcal{T}$ can be selected as follows:

$$\mathcal{T} \leftarrow \left\{(x, f) \in \mathcal{S} \mid \Delta_x^2 \leq r^2\right\} \quad [6.30]$$

where $r^2 = max(\chi_n^2(p), n_k)$. Finally, the problem has been circumvented and Algorithm 2 presents the KA-CMA-ES algorithm.

Algorithm 2: Kriging-assisted CMA-ES (KA-CMA-ES)

1: **Initialize** the evolution path $p_\sigma^{(0)} = 0$, $p_c^{(0)} = 0$, the covariance matrix $C^{(0)} = I$, the step size $\sigma^{(0)}$ and selection parameters according to [ELH 11].

2: **Initialize** the mean vector $m^{(0)}$ to a random candidate

3: $g \leftarrow 0$

4: **As long as** the convergence criterion is not satisfied do

5: $\quad x_k^{(g+1)} = m^{(g)} + \sigma^{(g)} \mathcal{N}(0, C^{(g)})$ for $k = 1, ..., \lambda$

$$T \leftarrow \left\{ (x, f) \in \mathcal{S} \mid \left(x - m^{(g)}\right)^T \left(\left(\sigma^{(g)}\right)^2 C^{(g)}\right)^{-1} \left(x - m^{(g)}\right) \leq r^2 \right\}$$

Si $|T| < n_k$ **then**

→ **Evaluation by the original function**

$$f_k \leftarrow f\left(x_k^{(g+1)}\right) \quad k = 1, \ldots, \lambda$$

$$\mathcal{S} = \mathcal{S} \cup \left\{\left(x_k^{(g+1)}, f_k\right)\right\}_{k=1}^{\lambda}$$

$$T \leftarrow \left\{ (x, f) \in \mathcal{S} \mid \left(x - m^{(g)}\right)^T \left(\left(\sigma^{(g)}\right)^2 C^{(g)}\right)^{-1} \left(x - m^{(g)}\right) \leq r^2 \right\}$$

Otherwise (the site of the parent individuals remains unchanged)

→ **Metamodel-based evaluation**

Perform the approximate ranking procedure according to **Algorithm 1}**

End if

$$m^{(g+1)} = \sum_{i=1}^{\mu} w_i x_{i:\lambda}^{(g)}$$

$$p_\sigma^{(g+1)} = (1 - c_\sigma) p_\sigma^{(g)} + \sqrt{c_\sigma (2 - c_\sigma) \mu_{eff}} \, (C^{(g)})^{-\frac{1}{2}} \frac{m^{(g+1)} - m^{(g)}}{\sigma^{(g)}}$$

$$\sigma^{(g+1)} = \sigma^{(g)} exp\left(\frac{c_\sigma}{d_\sigma}\left(\frac{\| p_\sigma^{(g+1)} \|}{E(\| \mathcal{N}(0, I) \|)} - 1\right)\right)$$

$$p_c^{(g+1)} = (1 - c_c) p_c^{(g)} + h_\sigma \sqrt{c_c (2 - c_c) \mu_{eff}} \frac{m^{(g+1)} - m^g}{\sigma^g}$$

$$C^{(g+1)} = (1 - c_1 - c_u + c_1 \delta(h_\sigma)) C^{(g)} + c_1 p_c^{(g+1)} (p_c^{(g+1)})^T + c_\mu \sum_{i=1}^{u} w_i \left(\frac{x_{i:\lambda}^{(g+1)} - m^{(g)}}{\sigma^{(g)}}\right) \ldots$$

$$\left(\frac{x_{i:\lambda}^{(g+1)} - m^{(g)}}{\sigma^{(g)}}\right)^T$$

End as long as

Return the best individual $\mathbf{x}_{\mathbf{best}}$

End for

6.6. Global optimization of the PQFP housing solder joints

6.6.1. *Formulation of the problem*

The design optimization problem is to maximize the number of fatigue lifecycles of the solder joints:

$$\max N_f(\{d\}) \\ \text{s.c.}\left\{\{d\}^{Lb} \leq \{d\} \leq \{d\}^{Ub}\right. \quad [6.31]$$

where $\{d\} = \{h_1, h_2, h_3\}$ are the design variables (see Figure 6.13 and Table 6.5) that represent the three design parameters that have a major impact on the fatigue strength of the solder joint [ELH 13a] and N_f represents the number of fatigue lifecycles.

The number of N_f fatigue cycles is calculated through plastic deformation using the Coffin–Manson empirical model. In other words, maximizing N_f is equivalent to minimizing the plastic deformation $\Delta\varepsilon_{in}$ of the solder joint after one thermal cycle under active load. Therefore, the objective function represents the plastic deformation $\Delta\varepsilon_{in}$ of the solder joint due to the last thermal load cycle [HAM 19].

$$\min \Delta\varepsilon_{in}(\{d\}) \\ \text{s.c.}\left\{\{d\}^{Lb} \leq \{d\} \leq \{d\}^{Ub}\right. \quad [6.32]$$

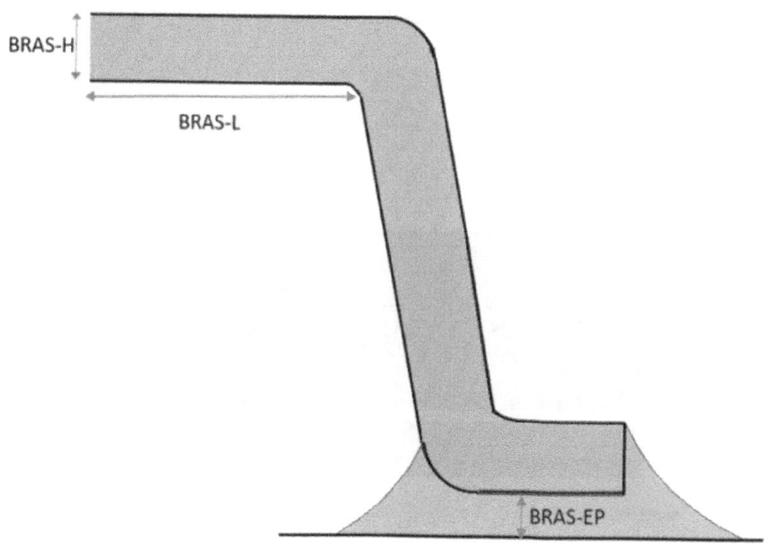

Figure 6.13. *Description of the optimization parameters*

Parameter (mm)	Lower limit $\{d\}^{Lb}$	Upper limit $\{d\}^{Lb}$
Young's modulus	0.1	0.25
Poisson's ratio	0.5	1
Density	0.1	0.4

Table 6.5. *Upper and lower limits of the optimization variables*

6.6.2. *Numerical simulations*

The design optimization problem is realized by the KA-CMA-ES algorithm, implemented on Matlab with ANSYS integration.

Table 6.6 presents the results of the optimal and initial design calculated by the simple CMA-ES algorithm and the KA-CMA-ES algorithm; also the results show that the objective function (change in non-plastic deformation $\Delta\varepsilon_{in}$) is minimized. Figure 6.15 shows the new design and the new inelastic strain distribution after applying the design optimization procedure. The maximum inelastic deformation (SMX = 0.08193) is less than the inelastic deformation of the initial configuration (SMX = 0.13557). Typical convergence histories for plastic deformation $\Delta\varepsilon_{in}$ are shown in Figure 6.16. These plots clearly indicate that the KA-CMA-ES converges faster than the standard CMA-ES and renders the same results [HAM 19a].

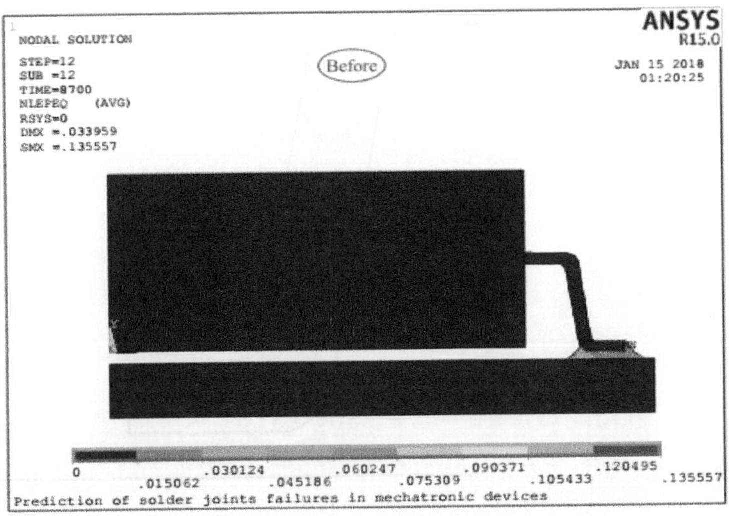

(a)

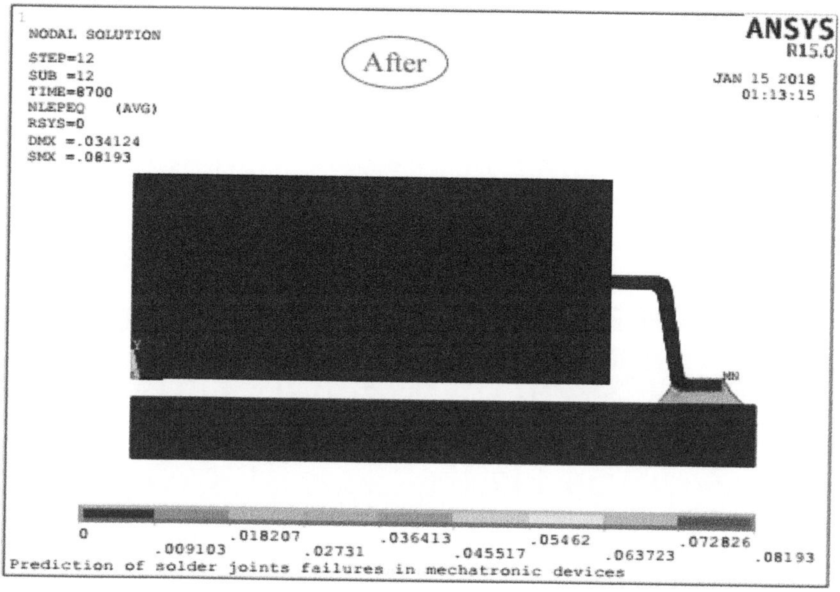

(b)

Figure 6.14. *Inelastic strain distribution before (a) and after (b) the optimization process*

Parameter (mm)	Initial value	CMA-ES	KA-CMA-ES
BRAS-EP	0.15	0.24	0.24
BRCH-L	0.7	0.83	0.83
BRCH-H	0.17	0.15	0.15
$\Delta\varepsilon_{in}$	0.025	0.0135	0.0135
N_f	940	2,185	2,185
Number of simulations	–	341	99
Duration of the simulations	–	61,380	17,820

Table 6.6. *Performance comparison of the two algorithms CMA-ES and KA-CMA-ES*

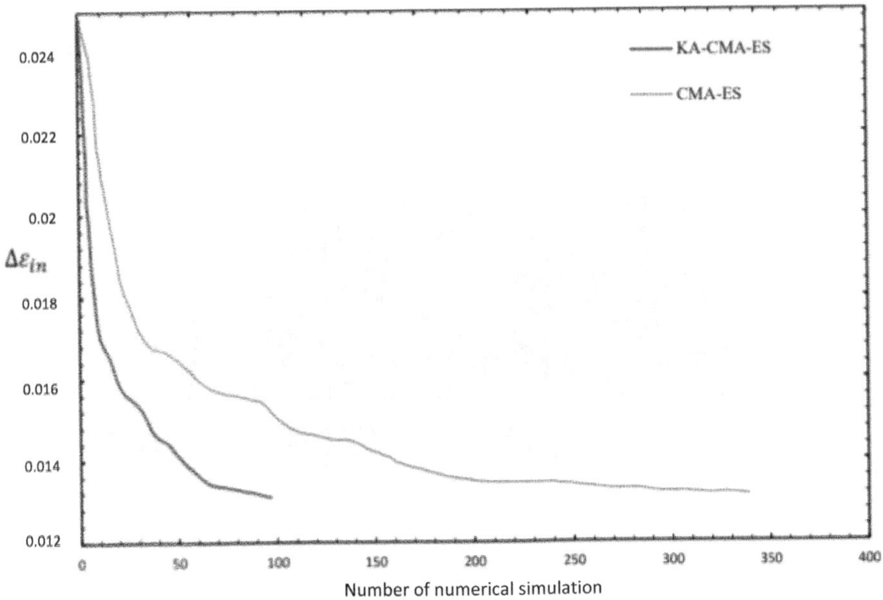

Figure 6.15. *Average curve of $\Delta\varepsilon_{in}$ as a function of the number of numerical simulations*

6.7. Conclusion

The robust design of complex mechatronic systems often induces an expensive optimization problem. In this chapter, the performance of the CMA-ES algorithm has been used with the Kriging metamodel to efficiently solve the problem of optimizing the solder joints of a PQFP package. The developed model takes into account the nonlinear properties of the viscoplastic behavior of solder joints. The study carried out illustrates the significance of using metamodeling techniques with CMA-ES in order to improve the performance of solder joints in mechatronic devices, while avoiding the issue of computational costs due to multiphysics simulation by FEM. The approximate ranking procedure, which is a reliable adaptive evolution control, was adopted to control the quality of the Kriging metamodel. Finally, the application of the KA-CMA-ES algorithm in the global solder joint optimization has shown its effectiveness in reducing the number of simulations per FEM as compared to the standard CMA-ES method.

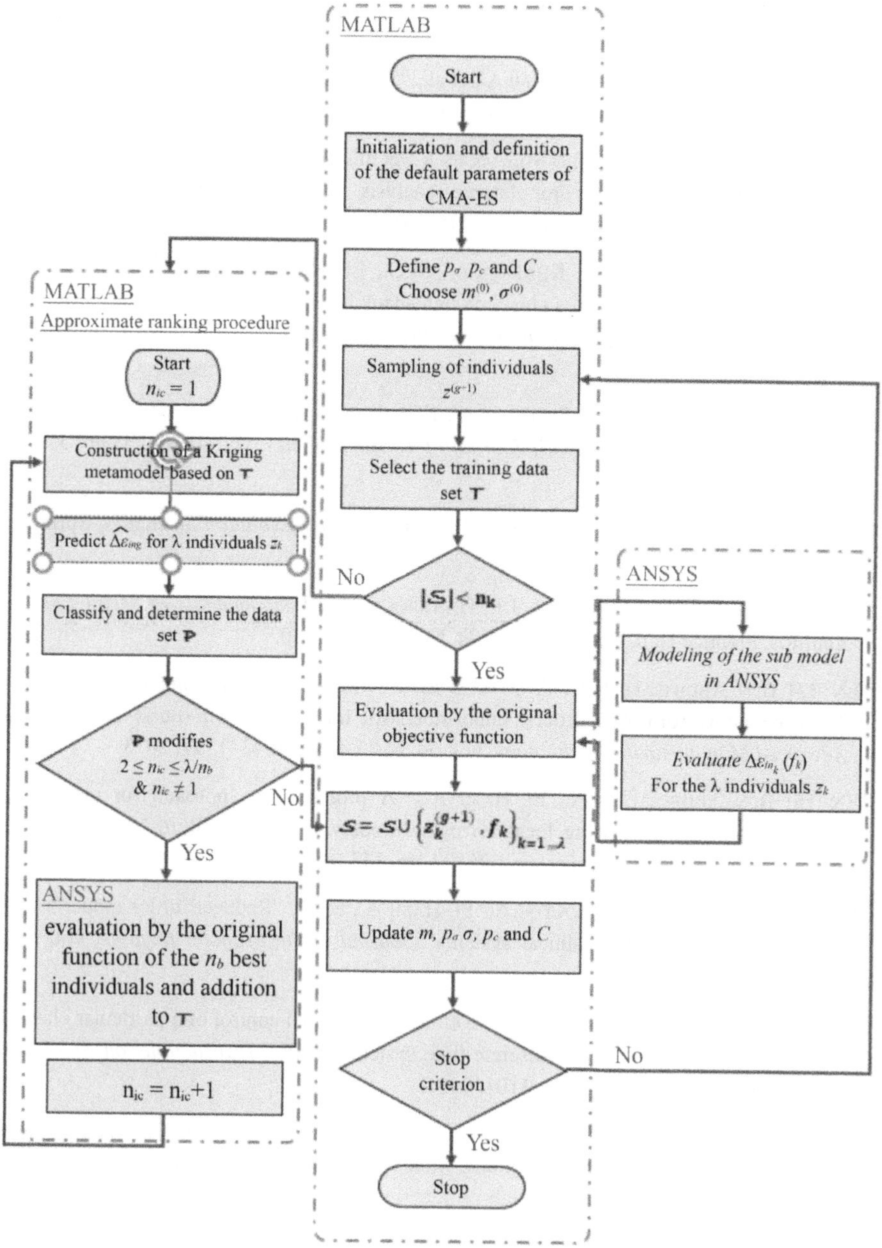

Figure 6.16. *Flow chart showing the global optimization methodology*

6.8. References

[ABI 20] ABID F., DAMMAK K., EL HAMI A., et al., "Surrogate models for uncertainty analysis of micro-actuator", *Microsystem Technologies*, pp. 1–12, 2020.

[ABO 11a] ABO AL-KHEER A., EID M., AOUES Y., et al., "Theoretical analysis of the spatial variability in tillage forces for fatigue analysis of tillage machines", *Journal of Terramechanics*, vol. 48, no. 4, pp. 285–295, 2011.

[ABO 11b] ABO AL-KHEER A., KHARMANDA M.G., EL HAMI A., et al., "Estimating the variability of tillage forces on a chisel plough shank by modeling the variability of tillage system parameters", *Computers and Electronics in Agriculture*, vol. 78, no. 1, pp. 61–70, 2011.

[ABO 11c] ABO AL-KHEER A., EL HAMI A., KHARMANDA M.G., et al., "Reliability-based design for soil tillage machines", *Journal of Terramechanics*, vol. 48, no. 1, pp. 57–64, 2011.

[BÄC 13] BÄCK T., FOUSSETTE C., KRAUSE P., *Contemporary Evolution Strategies*, Springer, Berlin, Heidelberg, 2013.

[BAN 90] BANNANTINE J.A., COMER J.J., HANDROCK J.L., *Fundamentals of Metal Fatigue Analysis*, Prentice Hall, Englewood Cliffs, New Jersey, 1990.

[BEN 13] BEN ABDESSALEM A., PAGNACCO E., EL HAMI A., "Increasing the stability of T-shape tube hydroforming process under stochastic framework", *International Journal of Advanced Manufacturing Technology*, vol. 69, nos 5–8, pp. 1343–1357, 2013.

[BEN 15] BEN ABDESSALEM A., EL HAMI A., "A probabilistic approach for optimising hydroformed structures using local surrogate models to control failures", *International Journal of Mechanical Sciences*, vols 96–97, pp. 143–162, 2015.

[BOU 96] BOUYEKHF R., EL MOUDNI A., EL HAMI A., et al., "Reduced order modelling of two-time-scale discrete non-linear systems", *Journal of the Franklin Institute*, vol. 333, no. 4, pp. 499–512, 1996.

[BOU 01] BOUYEKHF R., EL HAMI A., EL MOUDNI A., "Optimal control of a particular class of singularly perturbed nonlinear discrete-time systems", *IEEE Transactions on Automatic Control*, vol. 46, no. 7, pp. 1097–1101, 2001.

[BRA 05] BRANKE J., SCHMIDT C., "Faster convergence by means of fitness estimation", *Soft Computing – A Fusion of Foundations, Methodologies and Applications*, vol. 9, no. 1, pp. 13–20, 2005.

[COF 54] COFFIN Jr. L.F., "A study of the effects of cyclic thermal stresses on a ductile metal", *Transactions of the American Society of Mechanical Engineers*, vol. 76, no. 6, pp. 931–950, 1954.

[DAM 17] DAMMAK K., EL HAMI A., KOUBAA S., et al., "Reliability based design optimization of coupled acoustic-structure system using generalized polynomial chaos", *International Journal of Mechanical Sciences*, vol. 134, pp. 75–84, 2017.

[DAM 19a] DAMMAK K., KOUBAA S., EL HAMI A., et al., "Numerical modelling of vibro-acoustic problem in presence of uncertainty: Application to a vehicle cabin", *Applied Acoustics*, vol. 144, pp. 113–123, 2019.

[DAM 19b] DAMMAK K., EL HAMI A., "Multi-objective reliability-based design optimization of coupled acoustic-structural system", *Engineering Structures*, vol. 197, 109389, 2019.

[DAM 20a] DAMMAK K., YAICH A., EL HAMI A., et al., "An efficient optimization based on the robust hybrid method for the coupled acoustic–structural system", *Mechanics of Advanced Materials and Structures*, vol. 27, no. 21, pp. 1816–1826, 2020.

[DAM 20b] DAMMAK K., EL HAMI A., "Multi-objective reliability-based design optimization using Kriging surrogate model for cementless hip prosthesis", *Computer Methods in Biomechanics and Biomedical Engineering*, vol. 23, no. 12, pp. 854–865, 2020.

[DAM 21] DAMMAK K., EL HAMI A., "Thermal reliability-based design optimization using Kriging model of PCM based pin fin heat sink", *International Journal of Heat and Mass Transfer*, vol. 166, 120745, 2021.

[DAR 00] DARVEAUX R., "Effect of simulation methodology on solder joint crack growth correlation", *50th Proceedings on Electronic Components & Technology Conference*, IEEE, 2000.

[DEB 20] DEBICH B., EL HAMI A., YAICH A., et al., "An efficient reliability-based design optimization study for PCM-based heat-sink used for cooling electronic devices", *Mechanics of Advanced Materials and Structures*, vol. 29, no. 12, pp. 1661–1673, DOI:10.1080/15376494.2020.1836291, 2020.

[DEB 21] DEBICH B., Modélisation, simulation, et fiabilité des matériaux changement de phase pour l'automobile et l'aéronautique, Thesis, INSA Rouen Normandie, 2021.

[ELH 93] EL HAMI A., LALLEMENT G., MINOTTI P., et al., "Methods that combine finite group theory with component mode synthesis in the analysis of repetitive structures", *Computers and Structures*, vol. 48, no. 6, pp. 975–982, 1993.

[ELH 96] EL HAMI A., RADI B., "Some decomposition methods in the analysis of repetitive structures", *Computers and Structures*, vol. 58, no. 5, pp. 973–980, 1996.

[ELH 09] EL HAMI A., RADI B., CHEROUAT A., "Treatment of the composite fabric's shaping using a Lagrangian formulation", *Mathematical and Computer Modelling*, vol. 49, nos 7–8, pp. 1337–1349, 2009.

[ELH 11] EL HAMI A., RADI B., "Comparison study of different reliability-based design optimization approaches", *Advanced Materials Research*, vol. 274, pp. 113–121, 2011.

[ELH 13a] EL HAMI A., RADI B., *Incertitude et optimisation et fiabilité des structures*, Hermes-Lavoisier, Paris, 2013.

[ELH 13b] EL HAMI A., RADI B., *Uncertainty and Optimization in Structural Mechanics*, ISTE Ltd, London and John Wiley & Sons, New York, 2013.

[FAN 05] FANG K.-T., LI R., SUDJIANTO A., *Design and Modeling for Computer Experiments*, CRC Press, Boca Raton, Florida, 2005.

[GRI 10] GRIEU M., Étude de la fatigue des joints brasés de composants électroniques soumis à des sollicitations thermomécaniques, vibratoires et combinées, Thesis, Centre des Matériaux, Corbeil-Essonnes, 2010.

[GUE 15a] GUERINE A., EL HAMI A., WALHA L., et al., "Approach for the dynamic analysis of one stage gear system with uncertain parameters", *Mechanism and Machine Theory*, vol. 92, pp. 113–126, 2015.

[GUE 15b] GUERINE A., EL HAMI A., FAKHFAKH T., et al., "A polynomial chaos method to the analysis of the dynamic behavior of spur gear system", *Structural Engineering and Mechanics*, vol. 53, no. 4, pp. 819–831, 2015.

[GUE 16a] GUERINE A., EL HAMI A., WALHA L., et al., "A polynomial chaos method for the analysis of the dynamic behavior of uncertain gear friction system", *European Journal of Mechanics, A/Solids*, vol. 59, pp. 76–84, 2016.

[GUE 16b] GUERINE A., EL HAMI A., WALHA L., et al., "Dynamic response of a Spur gear system with uncertain friction coefficient", *Advances in Engineering Software*, vol. 120, pp. 45–54, 2016.

[GUE 17] GUERINE A., EL HAMI A., WALHA L., et al., "Dynamic response of wind turbine gear system with uncertain-but-bounded parameters using interval analysis method", *Renewable Energy*, vol. 113, pp. 679–687, 2017.

[HAM 17] HAMDANI H., RADI B., EL HAMI A., "Métamodélisation pour une conception robuste des systèmes mécatroniques", *Incertitudes et fiabilité des systèmes multiphysiques*, no. 2, November 2017.

[HAM 19a] HAMDANI H., BOUCHAÏB, R., EL HAMI A., "Metamodel assisted evolution strategies for global optimization of solder joints reliability in embedded mechatronic devices", *Microsystem Technologies*, vol. 25, no. 10, pp. 3801–3812, 2019.

[HAM 19b] HAMDANI H., BOUCHAÏB R., EL HAMI A., "Optimization of solder joints in embedded mechatronic systems via Kriging-assisted CMA-ES algorithm", *International Journal for Simulation and Multidisciplinary Design Optimization*, vol. 10, A3, 2019.

[HAN 15] HANSEN N., ARNOLD D.V., AUGER A., "Evolution strategies", in KACPRZYK, J., PEDRYCZ, W. (eds), *Springer Handbook of Computational Intelligence*, Springer, Berlin, Heidelberg, 2015.

[HAN 16] HANSEN N., The CMA evolution strategy: A tutorial, arXiv preprint, arXiv:1604.00772, 2016.

[HOS 07] HOSSAIN M.M., JAGARKAL S.G., AGONAFER D., et al., "Design optimization and reliability of PWB level electronic package", *Journal of Electronic Packaging*, vol. 129, no. 1, pp. 9–18, 2007.

[HUA 16] HUANG C., RADI B., EL HAMI A., "Uncertainty analysis of deep drawing using surrogate model based probabilistic method", *Manufacturing Technology*, vol. 86, nos 9–12, pp. 3229–3240, 2016.

[HUA 17] HUANG C., EL HAMI A., RADI B., "Metamodel-based inverse method for parameter identification: Elastic–plastic damage model", *Engineering Optimization*, vol. 49, no. 4, pp. 633–653, 2017.

[HUA 18] HUANG C., RADI B., EL HAMI A., et al., "CMA evolution strategy assisted by Kriging model and approximate ranking", *Applied Intelligence*, pp. 1–17, 2018.

[JAN 17] JANNOUN M., AOUES Y., PAGNACCO E., et al., "Probabilistic fatigue damage estimation of embedded electronic solder joints under random vibration", *Microelectronics Reliability*, vol. 78, pp. 249–257, 2017.

[JES 00] JESD22-A104-B, Temperature cycling, Jedec Solid State Technology Association, 2000.

[JIN 03] JIN Y., HÜSKEN M., SENDHOFF B., "Quality measures for approximate models in evolutionary computation", *GECCO*, pp. 170–173, 2003.

[JIN 05] JIN Y., "A comprehensive survey of fitness approximation in evolutionary computation", *Soft Computing – A Fusion of Foundations, Methodologies and Applications*, vol. 9, no. 1, pp. 3–12, 2005.

[JIN 11] JIN Y., "Surrogate-assisted evolutionary computation: Recent advances and future challenges", *Swarm and Evolutionary Computation*, vol. 1, no. 2, pp. 61–70, 2011.

[KAM 20] KAMEL A., DAMMAK K., EL HAMI A., et al., "A modified hybrid method for a reliability-based design optimization applied to an offshore wind turbine", *Mechanics of Advanced Materials and Structures*, vol. 29, no. 9, pp. 1229–1242, DOI:10.1080/15376494.2020.1811927, 2020.

[KAN 02] KANCHANOMAI C., MIYASHITA Y., MUTOH Y., "Low cycle fatigue behavior and mechanisms of a eutectic Sn–Pb solder 63Sn/37Pb", *International Journal of Fatigue*, vol. 24, no. 6, pp. 671–683, 2002.

[KER 06] KERN S., HANSEN N., KOUMOUTSAKOS P., "Local meta-models for optimization using evolution strategies", *PPSN*, vol. 4193, pp. 939–948, 2006.

[KHA 08] KHARMANDA G., MOHSINE A., MAKLOUFI A., et al. "Recent methodologies for reliability-based design optimization", *International Journal for Simulation and Multidisciplinary Design Optimization*, vol. 2, no. 1, pp. 11–23, 2008.

[KHA 09] KHARMANDA G., SHARABATEY S., IBRAHIM H., et al., "Reliability-based design optimization using semi-numerical strategies for structural engineering applications", *International Journal of CAD/CAM*, vol. 9, no. 1, pp. 1–16, 2009.

[KRA 14] KRAMER O., *A Brief Introduction to Continuous Evolutionary Optimization*, Springer, Cham, 2014.

[KRI 51] KRIGE D.G., "A statistical approach to some basic mine valuation problems on the Witwatersrand", *Journal of the Southern African Institute of Mining and Metallurgy*, vol. 52, no. 6, pp. 119–139, 1951.

[LED 17a] LEDOUX M., EL HAMI A., *Fluid Mechanics: Analytical Methods*, ISTE Ltd, London and John Wiley & Sons, New York, 2017.

[LED 17b] LEDOUX M., EL HAMI A., *Compressible Flow, Propulsion and Digital Approaches in Fluid Dynamics*, ISTE Ltd, London and John Wiley & Sons, New York, 2017.

[LED 21a] LEDOUX M., EL HAMI A., *Heat Transfer 1. Conduction*, ISTE Ltd, London and John Wiley & Sons, New York, 2021.

[LED 21b] LEDOUX M., EL HAMI A., *Heat Transfer 2. Radiative Transfer*, ISTE Ltd, London and John Wiley & Sons, New York, 2021.

[LIU 04] LIU D.S., CHAO Y.C., WANG C.H., "Study of wire bonding looping formation in the electronic packaging process using the three-dimensional finite element method", *Finite Elements in Analysis and Design*, vol. 40, no. 3, pp. 263–286, 2004.

[MAB 17] MABROUK I.B., EL HAMI A., WALHA L., et al., "Dynamic vibrations in wind energy systems: Application to vertical axis wind turbine", *Mechanical Systems and Signal Processing*, vol. 85, pp. 396–414, 2017.

[MAK 16] MAKHLOUFI A., AOUES Y., EL HAMI A., "Reliability based design optimization of wire bonding in power microelectronic devices", *Microsystem Technologies*, vol. 22, no. 12, pp. 2737–2748, 2016.

[MAT 63] MATHERON G., "Principles of geostatistics", *Economic Geology*, vol. 58, no. 8, pp. 1246–1266, 1963.

[MOH 10] MOHSINE A., EL HAMI A., "A robust study of reliability-based optimization methods under eigen-frequency", *Computer Methods in Applied Mechanics and Engineering*, vol. 199, nos 17–20, pp. 1006–1018, 2010.

[MOR 65] MORROW J., "Cyclic plastic strain energy and fatigue of metals", *Internal Friction, Damping, and Cyclic Plasticity*, ASTM International, 1965.

[MOR 02] MORO T., EL HAMI A., EL MOUDNI A., "Reliability analysis of a mechanical contact between deformable solids", *Probabilistic Engineering Mechanics*, vol. 17, no. 3, pp. 227–232, 2002.

[NUB 11] NUBLI ZULKIFLI M., AZHAR ZAHID JAMAL Z., ABDUL QUADIR G., "Temperature cycling analysis for ball grid array package using finite element analysis", *Microelectronics International*, vol. 28, no. 1, pp. 17–28, 2011.

[PAN 04] PANG J.H.L., XIONG B.S., LOW T.H., "Low cycle fatigue models for lead-free solders", *Thin Solid Films*, vols 462–463, pp. 408–412, 2004.

[POU 16] POUPON G., "Procédés de packaging et d'interconnexion de composants électroniques", *Techniques de l'ingénieur. Cartes électroniques : technologies et conception*, e3401, 2016.

[RAD 07] RADI B., EL HAMI A., "Reliability analysis of the metal forming process", *Mathematical and Computer Modelling*, vol. 45, nos 3–4, pp. 431–439, 2007.

[RAS 06] RASMUSSEN C.E., WILLIAMS C.K.I., *Gaussian Processes for Machine Learning* MIT Press, Cambridge, Massachusetts, 2006.

[REN 03] RENCHER A.C., *Methods of Multivariate Analysis*, John Wiley & Sons, New York, 2003.

[RUN 04] RUNARSSON T.P., "Constrained evolutionary optimization by approximate ranking and surrogate models", *International Conference on Parallel Problem Solving from Nature*, vol. 3242, pp. 401–410, 2004.

[SAC 89] SACKS J., WELCH W.J., MITCHELL T.J., et al., "Design and analysis of computer experiments", *Statistical Science*, vol. 4, no. 4, pp. 409–435, 1989.

[SAR 11] SARSRI D., AZRAR L., JEBBOURI A., et al., "Component mode synthesis and polynomial chaos expansions for stochastic frequency functions of large linear FE models", *Computers and Structures*, vol. 89, nos 3–4, pp. 346–356, 2011.

[SHI 00] SHI X.Q., PANG H.L.J., ZHOU W., et al., "Low cycle fatigue analysis of temperature and frequency effects in eutectic solder alloy", *International Journal of Fatigue*, vol. 22, no. 3, pp. 217–228, 2000.

[VAN 06] VAN DRIEL W., HOCHSTENBACH H., ZHANG G., "Design for reliability of wafer level packages", *Thermal, Mechanical and Multiphysics Simulation and Experiments in Micro-Electronics and Micro-Systems, 2006. 7th International Conference on EuroSime 2006*, IEEE, pp. 1–6, 2006.

[YAI 18] YAICH A., KHARMANDA G., EL HAMI A., et al., "Reliability based design optimization for multiaxial fatigue damage analysis using robust hybrid method", *Journal of Mechanics*, vol. 34, no. 5, pp. 551–566, 2018.

7

Reliable Optimization of Vibro-acoustic Problems in the Presence of Uncertainties via Polynomial Chaos

7.1. Introduction

Structural optimization has been of great interest to designers for several years. It is progressively integrated into the classical modeling techniques as its performance increases. At the beginning, optimization was used in very simple systems, but now its field of application is extending to new and more complex challenges.

Determining the optimal design of a system is a major challenge for designers. The ideal (optimal) design must guarantee several criteria such as reliability, durability and feasibility all the while minimizing the cost of the structure. Therefore, the objective of structural optimization is to find the best design that simultaneously reduces costs and increases performance.

Generally, the optimization is formulated by a problem of minimizing a function, called the “objective function” under constraints called “limitations”. This problem is usually solved using iterative algorithms [DAM 19] that require the computation of the gradients of the objective function and the constraints.

The implementation of structure optimization [BRU 14] is shown in Figure 7.1. Based on an initial design, defined by design variables, the optimization aims to iteratively determine the optimal design by verifying the performance criteria.

Chapter written by Khalil DAMMAK and Abdelkhalak EL HAMI.

For a color version of all the figures in this chapter, see: http://www.iste.co.uk/elhami/applied1.zip.

Optimization that does not take into account structural uncertainties can lead to unpredictable deviations in actual and expected performances. Besides, the presence of a high risk of design failure does not meet the desired reliability. For this reason, several reliability optimization methods have been developed in order to design structures that satisfy both economic and reliable performances. These methods have been applied in different fields, such as civil engineering [KUR 02], mechatronics [MAK 16], vibro-acoustics [DAM 17], aerodynamics [DAS 11, ELH 13b], biomechanics [DAM 20] and many more.

In the first part of this chapter, probabilistic methods for the quantification and propagation of uncertainties are presented. The robust (or stochastic) approach is based on the fundamental concept called "polynomial chaos" (PC), using orthogonal functions to represent second order random processes, solutions of stochastic systems. The second part of this chapter is devoted to the implementation of different reliability-based design optimization (RBDO) methods to solve vibro-acoustic problems in the presence of uncertainties.

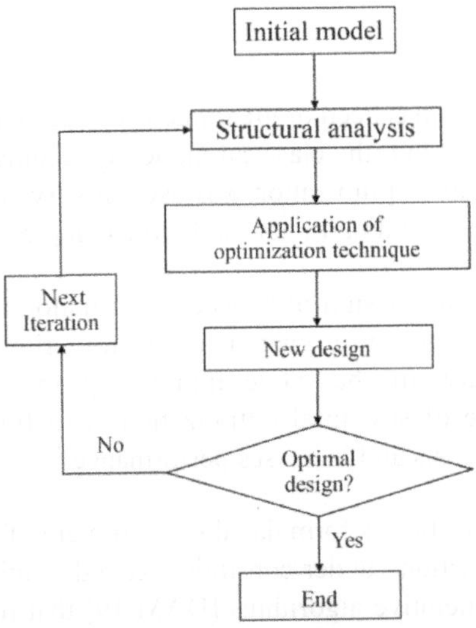

Figure 7.1. *Implementation of structure optimization*

7.2. Robust approaches to uncertainty propagation

7.2.1. *The Monte Carlo method*

As mentioned in Chapter 1, the Monte Carlo (MC) method is a very general mathematical tool that can be applied to all systems. It is a probabilistic approach allowing for the modeling of uncertain parameters in complex dynamic systems, such as vibro-acoustic problems. In this section, the fundamental algorithm of the MC technique can be summarized by the following five steps [DAM 19]:

– Step 1: probabilistic identification of uncertain parameters in the model.

– Step 2: sampling and random generation of realizations following the identified probabilistic laws.

– Step 3: propagation of uncertainty (the data sets generated by Step 2) in the model and the determination of the corresponding responses.

– Step 4: statistical analysis of the outputs corresponding to the samples.

– Step 5: analysis of the convergence in the distribution of the model output.

Let M be an analytical (or algorithmic) mathematical function modeling any system and $X = (X_1, X_2, .., X_r)$ a random vector of uncertain parameters. The response Y is the random solution of the following model:

$$Y = M(X) \quad [7.1]$$

Generally, the MC technique is used to estimate the distribution law of the output whose statistical characteristics are given by the mean value μ_Y and the standard deviation σ_Y. These two characteristics are calculated using a set of N simulations as follows:

$$\begin{cases} \hat{\mu}_Y = \frac{1}{N}\sum_{j=1}^{N} M(X^{(j)}) \\ \hat{\sigma}_Y^2 = \frac{1}{N-1}\sum_{j=1}^{N} (M(X^j) - \hat{\mu}_Y)^2 \end{cases} \quad [7.2]$$

According to the central limit theorem (CLT), the convergence of the MC technique is certified by the fact that the difference between the expectation μ_Y of Y and its estimate $\hat{\mu}_Y$ converges to a normal distribution with zero mean and a standard deviation equal to $\sigma_Y/\sqrt{N}$.

A rigorous use of the MC technique implies the need for confidence intervals among the results. For example, for a given number of draws N and a fixed confidence level $1 - \alpha$, the confidence interval corresponding to the mean value can be given as:

$$\hat{\mu}_Y - \frac{u_{\alpha/2}\hat{\sigma}_Y}{\sqrt{N}} \leq \mu \leq \hat{\mu}_Y + \frac{u_{\alpha/2}\hat{\sigma}_Y}{\sqrt{N}} \quad [7.3]$$

with $u_{\alpha/2} = -F_X^{-1}(1-\frac{\alpha}{2})$, F_X being the distribution function defined by $F(x) = \frac{1}{2\pi}\int_{-\infty}^{x} e^{-z^2}dz$. Equation [7.3] can be used to calculate the minimum number of draws, $N_{\min}$, needed to estimate the mean value with a relative error $\pm\varepsilon_{\mu_Y}$ with a confidence level equal to $1-\alpha$. This minimum number of draws can then be calculated by the following equation:

$$N_{\min} = 1 + ceil\left[\left(\frac{u_{\alpha/2}\hat{C}_Y}{\varepsilon_{\mu_Y}}\right)^2\right] \quad [7.4]$$

where $ceil(x)$ is the rounding of x to the nearest integer value and $\hat{C}_Y = \hat{\sigma}_Y/\hat{\mu}_Y$ represents the coefficient of variation for the response Y.

The simplicity of the MC technique has promoted its exploitation in different fields of engineering science. However, its major disadvantage lies in the fact that it is necessary to generate a large number of samples to have reasonable accuracy at a given level of confidence. Subsequently, the computational cost is highly prohibitive, especially for complex systems with a large number of degrees of freedom and/or many uncertain parameters.

7.2.2. *Generalized polynomial chaos*

7.2.2.1. *Overview*

The PC technique is a spectral representation based on the projection of random quantities in a probabilized space. This method is based on Wiener's homogeneous chaos theory and Cameron–Martin's theorem according to which a second-order stochastic process can be expressed by a convergent development, in the sense of least squares, in a series of polynomials, orthogonal functions of independent Gaussian random variables [DAM 20, WAN 17, MUK 17].

Theoretically, PC is a set of polynomial functions of the infinite set $\{\xi_i\}$ of random variables. In practice, the PC expansion is truncated to a finite number of terms:

$$X(\xi) = \sum_{j=0}^{N_p} \bar{x}_j \phi_j(\xi) \quad [7.5]$$

The number of terms $N_p + 1$ depends on the order of the GPC p and the number of uncertain parameters r, with $\xi = (\xi_1, \xi_2, \ldots, \xi_r)$. The relationship that gives the number of terms $N_p + 1$ as a function of p and r is expressed as follows:

$$N_p + 1 = \frac{(p+r)!}{p!r!} \quad [7.6]$$

The set $\{\phi_j\}$ groups polynomial functions that are orthogonal with respect to the Gaussian probability measure. The family ϕ_j is an orthogonal basis verifying the following orthogonality condition:

$$\langle \phi_i, \phi_j \rangle = \langle \phi_i, \phi_i \rangle \delta_{ij},\ \delta_{ij} = \begin{cases} 0 \text{ if } i \neq j \\ 1 \text{ if } i = j \end{cases} \quad [7.7]$$

δ_{ij} being the Kronecker symbol. The scalar product $\langle \phi_i, \phi_i \rangle$ is defined by:

$$\langle \phi_i, \phi_i \rangle = \int \phi_i \phi_i W(\xi) d\xi \quad [7.8]$$

The integration defined by the expression in equation [7.8] is performed over a domain determined by the probabilistic support of the random vector $\xi \in \mathbb{R}$, where $W(\xi)$ is the probability density function (PDF) of ξ described by:

$$W(\xi) = \prod_{l=1}^{r} \frac{\exp(-\xi_l^2/2)}{\sqrt{2\pi}} \quad [7.9]$$

The development [7.5] formalizes a separation between the stochastic and deterministic character of a random variable $X(\xi)$. The deterministic character is modeled by the stochastic coefficients $\bar{x}_j$ also called the modal coefficient. They can be static or dynamic (depending on the nature of the approximated random function) and represent the coordinates of the random variable $X(\xi)$ in its orthogonal basis of functions $\phi_j(\xi)$. The stochastic character of the variable $X(\xi)$ is concentrated in the orthogonal basis. The mode $\bar{x}_0$ of the PC development is just the mean value of the random variable $X(\xi)$ while the higher order modes provide a measure of the variability of $X(\xi)$ around its mean value. Indeed, we can directly determine its variance by the following equation:

$$\sigma_X^2 = \sum_{j=1}^{np} \bar{x}_j \langle \phi_j, \phi_j \rangle \quad [7.10]$$

Random functions can be described precisely by the Wiener–Hermite expansion. Its convergence, proved by the Cameron–Martin theorem, is verified in many practical examples. On the other hand, in the case of non-Gaussian functions, the speed and accuracy of convergence are particularly poor. In this context, an optimal correspondence, established by [MUK 17], generalizes the notion of PC to the non-Gaussian case. This generalization leads to the notion of GPC, also called the Wiener–Askey development because of the global correspondence, as defined in the Askey scheme [DAM 20].

	Random variable ξ	Wiener–Askey PC $\phi(\xi)$	Support
Continuous distribution	Gaussian	Hermite	$]-\infty,+\infty[$
	Uniform	Legendre	$[a,b]$
	Gamma	Laguerre	$[0,+\infty[$
	Beta	Jacobi	$[a,b]$

Table 7.1. *Correspondence between the type of distribution and the type of the polynomial chaos basis*

7.2.2.2. Practical computation in the development of polynomials for GPC

GPC allows for the quantification and modeling of uncertainties in the model based on three fundamental steps [DAM 21]:

– The choice of the polynomial basis appropriate to the problem:

GPC allows us to optimally represent a random function through a series development of orthogonal polynomials directly related to the probability law governing it (Askey scheme), which is usually not known. In practice, a test must first be performed with the MC method with a moderate number of samples so as to estimate the distribution law of the random function.

– The choice of the truncation of order N_p:

The choice of the chaos order p is essential for the accuracy of the approximation obtained with GPC. Generally, this parameter p is fixed via a convergence study that consists of evaluating the quality of the representation with the increase in the order.

– The computation of stochastic coefficients:

Computing a GPC representation requires the determination of $N_p + 1$ stochastic modes. Two implementation schemes can be used: the intrusive approach and the non-intrusive approach. The first one uses Galerkin projection techniques to express the uncertain model in the PC basis as a system of coupled and deterministic equations keeping the same linearity properties as the original system [ELH 13b, DEB 21]. The calculation of stochastic modes thus requires the adaptation of a calculation algorithm corresponding to the nature of the resulting system of equations. The consequence of this approach is the necessity for a single calculation to determine the stochastic coefficients. Nevertheless, in the case of systems with a large number of degrees of freedom, nonlinear systems and where the number of uncertain parameters is significant, this approach becomes very prohibitive in the sense of computational costs. All these drawbacks make the non-intrusive approach very appealing. Indeed, it does not require any modification or manipulation of the initial uncertain model for the calculation of the stochastic coefficients. The non-intrusive approach consists of projecting directly but also only the stochastic solution on the basis of the GPC. The difficulty is mainly related to obtaining the simulation points necessary to compute the stochastic coefficients by numerical integration methods.

In the following, only the non-intrusive approach is used in the thesis. Consider the problem of characterizing the solution $y(x,t,\xi(\omega))$ of the stochastic differential equation given in its following general form:

$$\Im(y(x,t,\xi(\omega)) = f(y(x,t,\xi(\omega)),x,\xi(\omega)) \qquad [7.11]$$

where $\Im$ denotes a differential operator. The solution $y(x,t,\xi(\omega))$ depends on the space x, time t and the random phenomenon modeled by the variable ξ, which itself is related to the random event ω. f is a function that can be nonlinear. The solution $y(x,t,\xi(\omega))$ can be expressed as a series expansion of orthogonal polynomials (equation [7.12]).

$$y(x,t,\xi) = \sum_{j=0}^{Np} \bar{y}_j(x,t)\phi_j(\xi) \qquad [7.12]$$

7.2.2.2.1. Non-intrusive approach

The major advantage of the non-intrusive approach lies in the fact that it requires a few calculations of deterministic solutions at a finite number of points in the probabilistic variation domain. The choice of these points depends essentially on the method used to compute the stochastic coefficients. Two techniques can be distinguished, namely the non-intrusive spectral projection (NISP) method and the regression method.

– NISP technique

In the NISP technique, once the solution is expressed by the GPC basis, according to the expression (equation [7.12]), it is projected onto it, which allows us to directly determine the stochastic coefficients as per equation [7.13]:

$$\bar{y}_l(x,t) = \frac{\langle y(x,t,\xi),\phi_l(\xi)\rangle}{\langle \phi_l(\xi),\phi_l(\xi)\rangle} \quad ,l = 0,..,N_p \qquad [7.13]$$

– Or the equivalent:

$$\bar{y}_l(x,t) = \frac{1}{\langle \phi_l{}^2 \rangle} \int y(x,t,\xi)\phi_l(\xi)W(\xi)d\xi \quad ,l = 0,..,N_p \qquad [7.14]$$

The coefficient $\bar{y}_l(x,t)$ measures, within one normalization factor, the correlation between the function $y(x,t,\xi)$ and the polynomial $\phi_l(\xi)$. The denominator in equation [7.14] is defined by inner products between identical polynomials. These inner products can be computed independently. Their values, which are real constants, can be tabulated and saved. The computation of the numerator in equation [7.14] involves the calculation of multidimensional integrals along the probabilistic support of the random vector ξ. Different methods are used to approximate these integrals such as numerical MC techniques, Simpson's method or Gauss collocation methods. For example, in the case of a one-dimensional Gauss integration, the approximation of the integral (equation [7.14]), is given by:

$$\bar{y}_l(x,t) = \frac{1}{\langle \phi_l{}^2 \rangle} \sum_{k=1}^{Q} y(x,t,\xi^{(k)})\phi_l(\xi^{(k)})W(\xi^{(k)}) \quad ,l = 0,..,N_p \qquad [7.15]$$

The points $\xi^{(k)}$ are the Gauss integration points also called collocation points and are given by the roots of the polynomial ϕ_l of degree p while the $W(\xi^{(k)})$ represent the Gauss integration weights.

– The regression technique:

The regression method consists of calculating the stochastic modes in such a way as to minimize, in a sense of least squares, the deviation ε between the solution of the stochastic model and its approximation in the GPC basis.

$$\varepsilon = \sum_{k=1}^{Q} \left[y(x,t,\xi^{(k)}) - \sum_{l=0}^{Np} \bar{y}_l(x,t)\phi_l(\xi^{(k)}) \right]^2 \qquad [7.16]$$

The regression method does not use the property of orthogonality for the polynomial basis, contrary to the NISP technique or the intrusive method. On the other hand, it requires a number of simulations Q greater than the number of modal coefficients to be calculated, hence $N_p + 1 < Q$. This condition is necessary for the solution of the corresponding least squares problem to be admissible. Q defines the number of simulations corresponding to a set of $\xi^{(k)}$ values. Two different approaches are possible to generate this set of values:

– the random approach whereby the $\xi^{(k)}$ points are generated according to the law ξ;

– the deterministic approach in which the points $\xi^{(k)}$ for the model are chosen in advance to be evaluated by equation [7.16]. The $\xi^{(k)}$ can be chosen from the roots of a polynomial with the condition $Q > N_p + 1$.

The vector of modal coefficients is denoted by $\bar{Y} = (\bar{y}_0, \ldots, \bar{y}_{Np})^T$, the element matrix Z by $z_{q,l} = \phi_l(\xi^{(q)})$ and the vector of simulations by $Y = (y(x,t,\xi^{(1)}), \ldots, y(x,t,\xi^{(q)}))$, corresponding to the set $\{\xi^{(q)}\}$, and if the matrix $Z^T Z$ is non-singular, then the optimal solution of the classical least squares problem is given as:

$$\bar{Y} = (Z^T Z)^{-1} Z^T Y \qquad [7.17]$$

The quality of the solution depends on the conditioning of the $Z^T Z$ matrix, called the Fischer matrix. Methods such as singular value decomposition can be used to further condition this matrix.

7.2.2.2.2. Criteria for selecting the optimal order of truncation

The higher the order of truncation, the better the accuracy of the GPC, but on the other hand, the higher the number of simulation points for the initial stochastic system. Therefore, the optimal order of truncation must be determined. The variance error $e_{\sigma^2,p}$ between two successive order processes is given as:

$$e_{\sigma^2,p} = \sigma_p^2 - \sigma_{p-1}^2 = \sum_{j=N_{p-1}+1}^{N_p} \bar{x}_j^2 \langle \phi_j^2 \rangle \qquad [7.18]$$

The average error (in the sense of least squares) between two successive order developments is calculated by:

$$\varepsilon_p^2 = \frac{1}{N'} \sum_{k=1}^{N'} \left(x_p(\xi_k) - x_{p-1}(\xi_k) \right)^2 = \frac{1}{N'} \sum_{k=1}^{N'} \left(\sum_{j=N_{p-1}+1}^{N_p} \bar{x}_j \phi_j(\xi_k) \right)^2 \qquad [7.19]$$

with N' the number of samples. We can show that this error is equal to:

$$\varepsilon_p^2 = e_{\sigma^2,p} \qquad [7.20]$$

It can be seen that it is not necessary to run simulations with the initial model or with the GPC to calculate the error that can be determined from the stochastic modes. The criterion for choosing the optimal order is based on the calculation of the rate of decrease in the relative error for the variance between two successive order developments:

$$\eta_i = \frac{e_{\sigma^2,i,p}}{\sigma^2{}_{i,p}} \qquad [7.21]$$

Therefore, we can write:

$$\sqrt{\eta_i} \leq \theta \qquad [7.22]$$

where θ is related to the threshold for choosing the truncation order [ELH 11]. Another criterion can be used to choose the optimal truncation order that is related to the number of direct computations $Q < N$ (number of Gauss collocation points), with $Q = (p+1)^r$ and N: the number of samples used by the MC method.

7.3. Structural optimization

7.3.1. *Formulation of the optimization problem*

The design optimization is formulated by the search for parameters intervening in the design phase, which minimize an objective function (in mechanics, the objective is often to minimize the mass of the structure or to maximize its stiffness) and which verify the mechanical, physical and geometrical performance constraints.

$$\begin{array}{ll} \min\limits_{x} & f(x) \\ \text{s.t.} & g_k(x) \leq 0, k = 1,..,K \\ & h_j(x) = 0, j = 1,..,J \\ & x_L \leq x \leq x_U \end{array} \qquad [7.23]$$

where x is the vector of design variables (also called decision variables), $f(x)$ is the objective function to be minimized, $g_k(x)$ and $h_j(x)$ are, respectively, the inequality and equality constraint, and x_L, x_U are, respectively, the lower and upper bounds of the search domain for the decision variables. It is possible to switch from a maximization problem to a minimization problem because of the following property:

$$\min_{x} \ f(x) = \max_{x} \ (-f(x)) \qquad [7.24]$$

Optimization algorithms can be classified into two classes: with and without gradient of the objective function and also constraints. In general, these algorithms converge to a local minimum rather than a global minimum. Figure 7.2 shows an example of a one-variable function with local minimums and a global minimum.

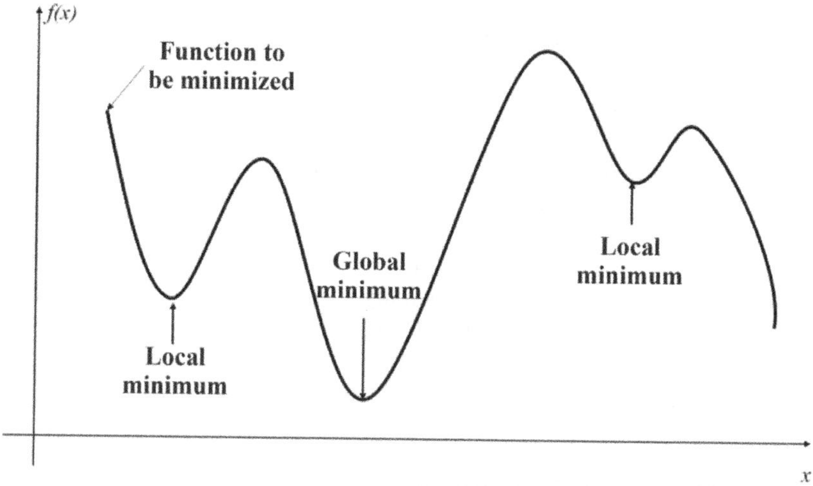

Figure 7.2. *Minimization of a one-variable objective function*

7.3.2. Deterministic design optimization

Deterministic design optimization (DDO) as formulated in equation [7.23] does not take into account the uncertainties of the input variables. In DDO, the designer introduces safety coefficients to increase the safety of the design. These coefficients are used to replace the presence of uncertainties in the structure. The choice of these safety factors is based on the engineers' experience.

The DDO process is illustrated in Figure 7.3. It follows three steps:

– representation of the geometry by a computer-assisted design (CAD) model;

– analysis of the model by the finite element method;

– optimization of the structure by minimizing an objective function such as volume.

Several algorithms and techniques have been developed to solve the DDO problem [KHA 08].

Although the DDO approach provides an optimal design, the reliability level cannot be controlled [DAM 21] since the safety coefficients are based on experience. Hence, the interest of reliability optimization to guarantee both the optimality and the reliability of the design.

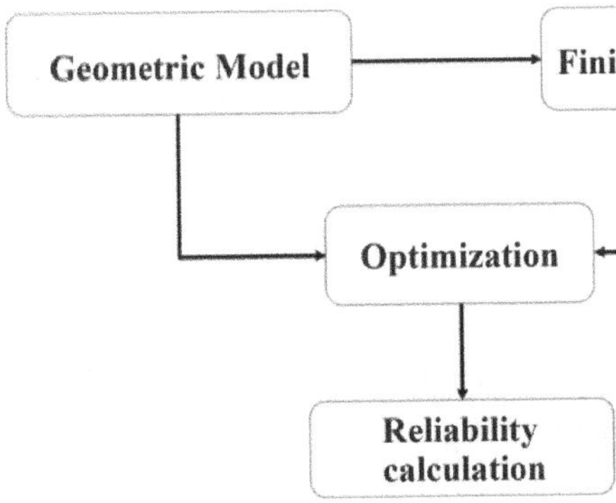

Figure 7.3. *Deterministic design optimization process*

7.3.3. *Reliability design optimization*

Reliability optimization is based on a failure analysis at each iteration in order to determine the roles of uncertainties on the system. However, in practical applications, the coupling between reliability analysis and optimization procedures leads to very high computation times and low convergence stability. In the RBDO process, as presented in Figure 7.4, two nested loops are presented: one is concerned with the optimization of the objective function under physical, geometrical, functional constraints, etc., and the other is devoted to the calculation of reliability, which itself requires an optimization procedure of the system performance function.

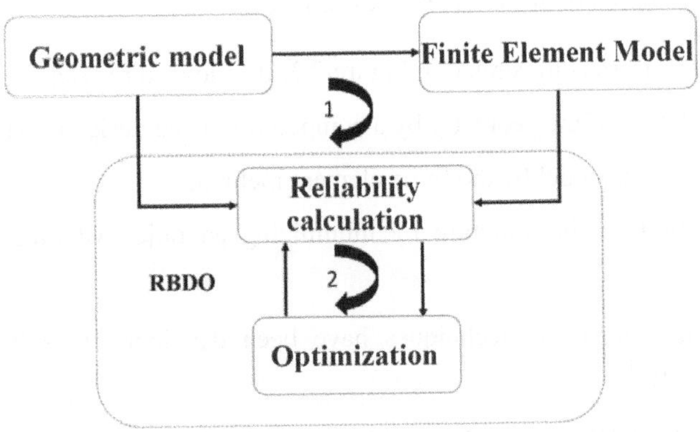

Figure 7.4. *The reliability-based design optimization process*

7.3.3.1. *Formulation of RBDO*

Unlike deterministic optimization, RBDO aims at finding the optimal solution, such that its performance function (failure limit state) is kept far enough from the design point.

The basic formulation of RBDO can be written as follows:

$$\begin{aligned} \min_{x} \quad & f(x) \\ \text{s.t.} \quad & P_r[G_i(x,y) \leq 0] \leq P_i^T, i = 1,..,m \\ & h_j(x) \leq 0, j = m+1,..,M \end{aligned} \qquad [7.25]$$

where x is the vector of deterministic design variables, y is the vector of random variables, f is the objective function (cost, mass, etc.), $P_r[.]$ is the probability operator, P_i^T is the target reliability level, $G_i(x,y)$ is the performance function (limit state) and $h_j(x)$ is the deterministic stress.

This formulation (equation [7.25]) is at the origin of all developments in RBDO. It can be realized by two types of algorithms:

– stochastic algorithms based on random simulations, such as genetic algorithms;

– nonlinear programming (NLP) algorithms based on the calculation of gradients of the objective function and mechanical constraints.

The computation of the reliability index requires a very significant computation time. In order to improve the numerical performance, Kharmanda et al. have developed an efficient method called the hybrid method (HM) [ELH 11]. It consists of solving simultaneously the reliability problem and the optimization problem. The solution is realized in a new space, called hybrid space design (HSD), containing both deterministic and random variables. This method has reduced the computation time enormously, however, the RBDO problem becomes more complex than the DDO problem. Therefore, another method called the optimum safety factor (OSF) was developed [ELH 11, KAR 09]. The optimal solution is obtained by considering the design point and the safety factors calculated from the derivatives of the constraints. This method reduced the computational time when compared to that of the HM. The efficiency of these methods has been demonstrated only for specific static and dynamic cases. An improvement of the HM called the robust hybrid method (RHM) is proposed by [YAI 18] and Dammak and El Hami [DAM 19] to cover the setbacks found in both the HM and the OSF methods.

7.3.3.2. *Classical approach*

Traditionally, classical RBDO consists, first of all, of performing a mechanical reliability analysis. The obtained solution is then used to minimize the objective function under mechanical constraints including the reliability constraint. As a result, the classical RBDO formulation is performed in two separate spaces (see Figure 7.5): the physical space and the normalized space [ELH 11] through the integration of the following two problems:

– *Optimization problem*

This problem consists of minimizing an objective function $f(x)$ under deterministic constraints $g_k(x)$ as well as reliability requirements. Its formulation can be defined as follows:

$$\begin{aligned} \min_{x} \quad & f(x) \\ \text{s.t.} \quad & g_k(x) \leq 0,\ k = 1,..,K \\ & \beta(x,u) \geq \beta_t \end{aligned} \qquad [7.26]$$

where u is the vector modulus in the normalized space, $\beta(x,u)$ is the reliability index of the structure and β_t is the desired reliability index.

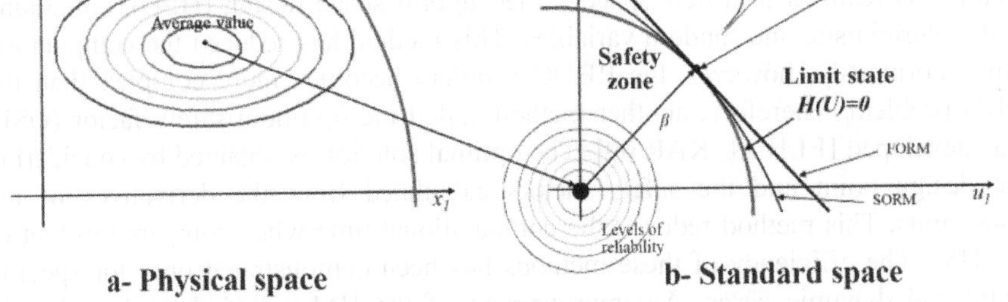

Figure 7.5. *Iso-probabilistic transformation of the physical space to the standard normal space*

– *Reliability problem*

The reliability index, which is taken as the minimum distance between the origin and the limit state function $H(x,u)$ in the normalized space, is determined by solving the following minimization problem:

$$\begin{aligned} \min_{u} \quad & d(u) = \sqrt{\sum u_i^2} \\ \text{s.t.} \quad & H(x,u) \leq 0 \end{aligned} \qquad [7.27]$$

The algorithm for this classical formulation is shown in Figure 7.6.

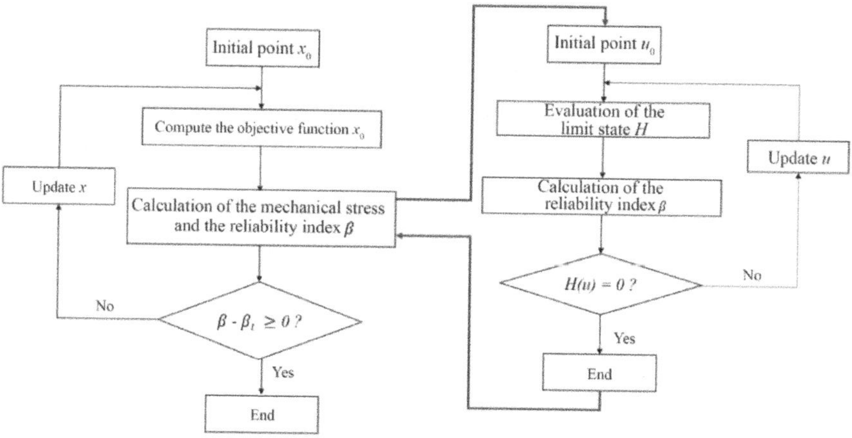

Figure 7.6. *Classical approach algorithm*

It is clear that in this classical approach, the total number of iterations of the two problems leads to a large number of evaluations of the mechanical model, and consequently, a sufficiently long computation time.

7.3.3.3. *Optimal safety factors*

In the DDO approach, the choice of the safety factor is based on engineering experience and the resulting solution may not satisfy the appropriate level of reliability. However, when designing a new structure, we usually need some experimental background to propose appropriate safety factors. However, with the OSF approach, the determination of these safety factors should be independent of the engineers' experience.

Although the HM leads to improvements for the reliability level, the OSF approach leads us to minimize the computation time. It consists of using sensitivity analysis combined with reliability analysis in order to eliminate the drawbacks of the DDO method and the RBDO methods [ELH 13a, KHA 08]. The algorithm of the OSF method consists of three main steps, as presented in Figure 7.7.

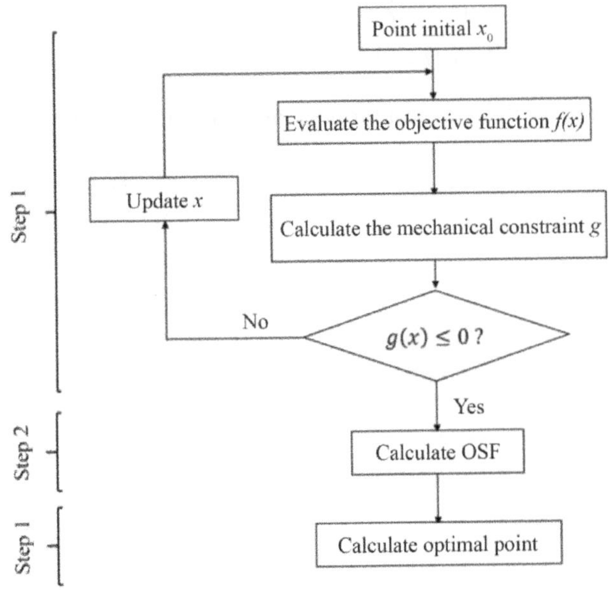

Figure 7.7. *Algorithm for the OSF method*

– *Determination of the design point* y_i^*

Here, we minimize the objective function under the conditions of the limit state function $G(y)$ and deterministic constraints. The resulting solution is designed by the most likely failure point.

– *Calculation of safety factors* S_f

To calculate these safety factors, a sensitivity analysis of the limit state function is required. In general, for a normal distribution of the random variable y_i, and for a problem with n design variables, the transformation from physical space to normalized space can be defined by equation [7.28]:

$$u_i = \frac{y_i - x_i}{\sigma_i} \quad , \; i = 1, \ldots, n \qquad [7.28]$$

Therefore, the safety factors can be determined from the u_i^* by:

$$S_{f_i} = 1 + \gamma_i . u_i^* \ , \ i = 1, \dots, n \qquad [7.29]$$

where γ_i is the coefficient relating the mean value and the standard deviation $\gamma_i = \frac{\sigma_i}{m_i}$.

According to El Hami and Radi [ELH 13a], the optimal value of the normalized vector u_i^* can be determined from the derivatives of the limit state function at the design point as follows:

$$u_i^* = \pm \beta_t \sqrt{\frac{|\frac{\partial G}{\partial y_i}|}{\sum_{i=1}^{n} |\frac{\partial G}{\partial y_i}|}}, i = 1, .., n \qquad [7.30]$$

The ± sign depends on the sign of the derivative:

$$\begin{cases} \frac{\partial G}{\partial y_i} > 0 \Leftrightarrow u_i^* > 0 \\ \frac{\partial G}{\partial y_i} < 0 \Leftrightarrow u_i^* < 0 \end{cases} , \ i = 1, .., n \qquad [7.31]$$

– *Calculation of the optimal solution* x_i^*

In the last step, to determine the optimal design, the values for the safety factors must be added into the design point value by the following equation:

$$x_i^* = y_i^* / S_{f_i} \qquad [7.32]$$

7.4. OSF method coupled with GPC applied to vibro-acoustic systems in the presence of uncertainties

In this section, a numerical example is studied: it is a flexible plate coupled with an acoustic cavity [DAM 17]. In this application, an analytical procedure is considered in the deterministic case. Next, a probabilistic analysis is implemented to study the robustness of the system. Finally, a reliability optimization using GPC is addressed to obtain the optimal robust system design.

Figure 7.8 shows a schematic of a rectangular cavity with five rigid walls coupled with a flexible plate simply supported on the end wall. A harmonic point force acts on the plate which provides the vibration. The vibrating plate thus generates an acoustic pressure inside the cavity.

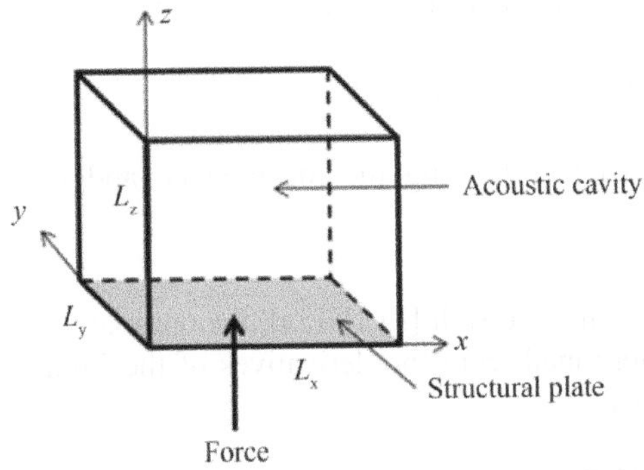

Figure 7.8. *Simply supported plate coupled to the end of a rectangular cavity with rigid walls*

The sizing and properties of the system are shown in Table 7.2.

Description	Parameters	Values
Width of the cavity	L_x	0.5 m
Cavity height	L_y	0.3 m
Cavity depth	L_z	1.1 m
Speed of sound	c	344 (m/s)
Air density	ρ	1.21 (kg/m^3)
Outlet location	–	(0.125 m, 0.15 m, 0.875 m)
Plate width	L_x	0.5 m
Plate height	L_y	0.3 m
Plate thickness	h	3 mm
Density of the plate	ρ_p	2,700 (kg/m^3)
The Young's modulus of the plate	E	70 (GPa)
The Poisson's ratio of the plate	ν	0.3
Amplitude of the force	F_z	1 (N)
Location of the force	–	(0.1 m, 0.075 m, 0 m)

Table 7.2. *Sizing and properties of the system*

The natural frequencies of a simply supported rectangular plate can be calculated by:

$$\omega_m(m_1, m_2) = \sqrt{\frac{D}{h.\rho_p}}.\left(\left(\frac{m_1.\pi}{L_x}\right)^2 + \left(\frac{m_2.\pi}{L_y}\right)^2\right) \quad [7.33]$$

where m_1, m_2 are the modal indices along the x, y axes, respectively, and D is the bending stiffness. The displacement of a point on the flexible plate can be described in terms of the summation of the structural modal participation factors u_m and its mode shapes ϕ_m as:

$$u(x, y) = \sum_{m=1}^{N_s} \phi_m(x, y).u_m, \quad [7.34]$$

where N_s is the number of structural modes and ϕ_m are given by:

$$\phi_{m_1,m_2}(x, y) = \sin\left(\frac{m_1.\pi.x}{L_x}\right).\sin\left(\frac{m_2.\pi.y}{L_y}\right) \quad [7.35]$$

The natural frequencies of a rectangular acoustic cavity with rigid walls are given by:

$$\omega_n(n_1, n_2, n_3) = \pi.c.\sqrt{\left(\frac{n_1}{L_x}\right)^2 + \left(\frac{n_2}{L_y}\right)^2 + \left(\frac{n_3}{L_z}\right)^2} \quad [7.36]$$

where n_1, n_2, n_3 are the modal indices along the x, y, z axes of the rectangular cavity. The acoustic pressure inside the cavity can be described in terms of the summation of the acoustic modal participation factors p_n and its mode shapes ψ_n through the form:

$$p(x, y, z) = \sum_{n=1}^{N_a} \psi_n(x, y, z).p_n, \quad [7.37]$$

where N_a is the number of acoustic modes and ψ_n, given as:

$$\psi_{n_1,n_2,n_3}(x, y, z) = \cos\left(\frac{n_1.\pi.x}{L_x}\right).\cos\left(\frac{n_2.\pi.y}{L_y}\right).\cos\left(\frac{n_3.\pi.y}{L_z}\right) \quad [7.38]$$

The motion equations of the plate and the acoustic response of the cavity can be rearranged into a matrix equation as shown in equation [7.39]:

$$\begin{bmatrix} \Lambda_m.(\omega_m^2 - \omega^2) & -S.[C_{nm}]^T \\ (-\omega^2).S.[C_{nm}] & \frac{\Lambda_n}{\rho.c^2}(\omega_n^2 - \omega^2) \end{bmatrix} \cdot \begin{pmatrix} u_m \\ p_n \end{pmatrix} = \begin{pmatrix} F_m \\ 0 \end{pmatrix} \quad [7.39]$$

where F_m is the modal force acting on the plate, S is the area of the structure, Λ_m is the modal mass of the simply supported plate, Λ_n is the modal volume of the rectangular cavity and C_{nm} is the dimensionless coupling coefficient between the structural mode, m, and the acoustic mode, n.

For the optimization problem, the objective function is to minimize the mass, W, of the plate subjected to the probabilistic stress related to the acoustic pressure. We consider the design points to be associated with the thickness of the plate and the magnitude of the force applied to the plate.

To achieve the DDO, a safety factor $S_f = 1.2$ is proposed for the target acoustic pressure $P_t = 50$ Pa.

The mathematical model of the DDO is formulated as follows:

$$\begin{aligned} &\min_{x} && W(x) \\ &\text{s.t.} && P_{max}(x,\xi) \leq \frac{P_t}{S_f} \end{aligned} \quad [7.40]$$

To perform RBDO using the OSF method, the first step is to obtain the design point. The formulation of the optimization problem is given by:

$$\begin{aligned} &\min_{y} && W(y) \\ &\text{s.t.} && P_{max}(y,\xi) \leq P_t = 50Pa \end{aligned} \quad [7.41]$$

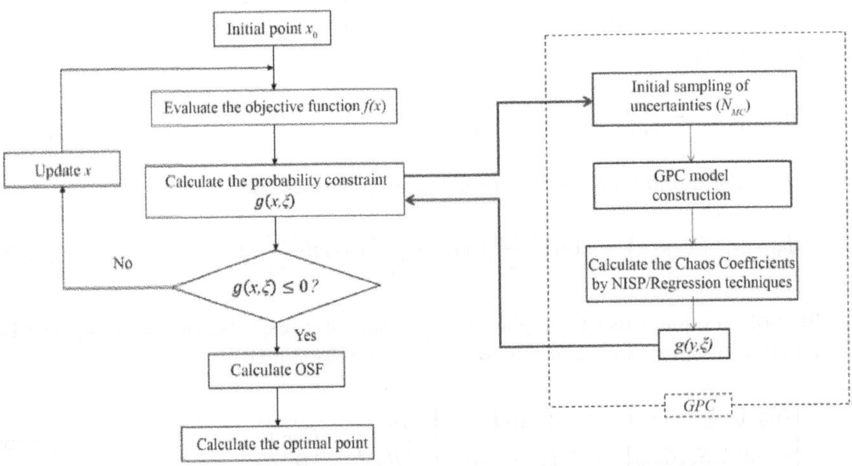

Figure 7.9. *Flowchart showing the OSF combined with the GPC*

The next step is to calculate the safety factors using equation [7.29] and determine the optimal solution by dividing the design point value by the safety coefficients calculated by equation [7.31]. The algorithm of the OSF method combined with GPC is shown in Figure 7.9.

The optimization problem is solved using the sequential quadratic programming (SQP) algorithm in the MATLAB software.

7.4.1. *Deterministic model*

The theoretical model described above has been implemented in MATLAB. As in equations [7.33] and [7.34], Tables 1.3 and 1.4, respectively, show the first four natural frequencies of the flexible plate and the acoustic cavity. A comparison with a finite element simulation using ANSYS for the uncoupled case is performed as a validation. It is remarkable that the theoretical formulation gives accurate results when compared with the finite element analysis.

Modes	Analytical (Hz)	ANSYS-FE (Hz)
1	110.43	108.30
2	198.12	196.36
3	344.27	344.80
4	354.01	359.96

Table 7.3. *Natural frequencies of the flexible plate*

Modes	Analytical (Hz)	ANSYS-FE (Hz)
1	156.36	159.53
2	312.73	313.51
3	344	347.40
4	377.87	378.96

Table 7.4. *Natural frequencies of the rectangular acoustic cavity*

Figure 7.10 shows the amplitude of the acoustic pressure in the cavity at the output location, and Figure 7.11 shows the amplitude of the plate displacement at the force location in a frequency band of [0 – 500]Hz. These results are compared with finite element simulations using ANSYS.

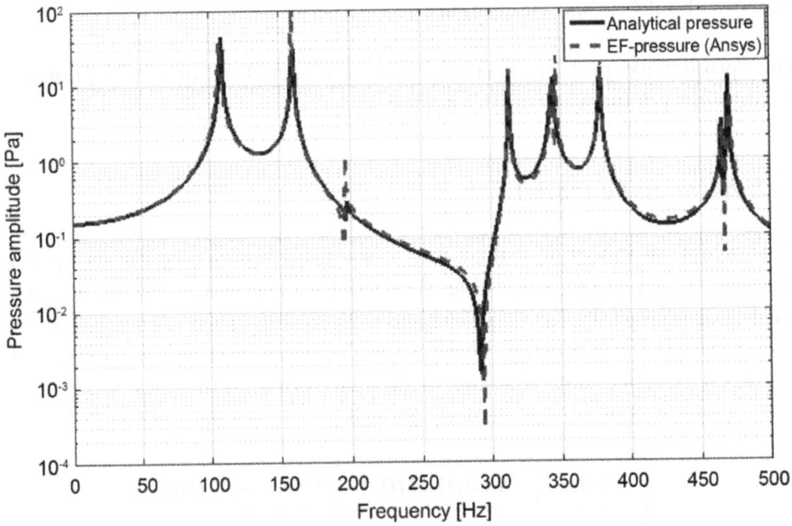

Figure 7.10. *Acoustic pressure inside the cavity*

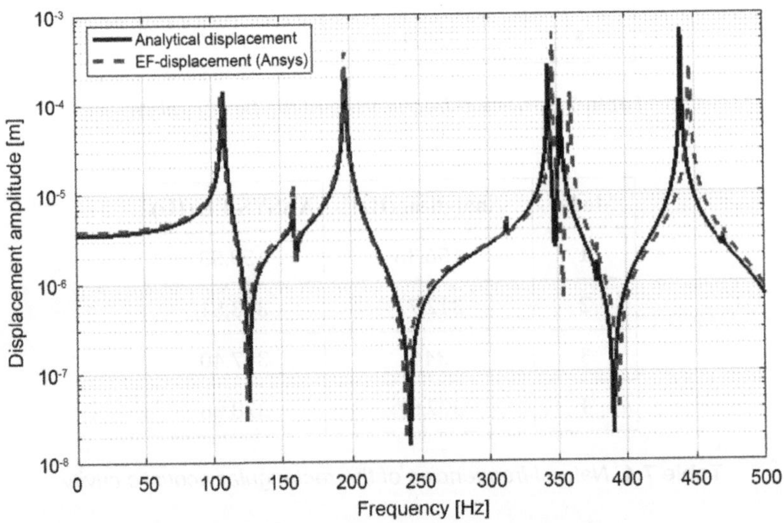

Figure 7.11. *Displacement of the plate at the location of the force*

It can be noted that the frequency peaks in the calculated results coincide with the resonant frequencies of the plate and the acoustic cavity, which were caused either by the modes of the cavity or by the modes of vibration on the plate. An excellent agreement between the analytical and numerical results is obtained which evaluates the model.

7.4.2. *Probabilistic analysis*

In this section, the stochastic method discussed above, GPC, is implemented and integrated into the analytical simulation described above. The objective is to improve the performance and robustness of coupled vibro-acoustic systems that operate in the presence of unavoidable uncertain parameters associated with material properties. These parameters are chosen randomly according to a uniform distribution around their nominal values $\pm 5\%$, as presented in Table 7.5.

	Acoustic part	Structural part	
Parameters	c (m/s)	E (GPa)	ν
Values	$U[326.8,361.2]$	$U[67.35,74.45]$	$U[0.285,0.315]$

Table 7.5. *Characteristics of uncertain parameters*

The use of the MC method to analyze the acoustic pressure inside the rectangular cavity and the plate displacement consists of creating a grid of numerical values from the uncertain parameters and calculating the quantity of interest for the linearized system for each value of this grid. The quantity of interest is analyzed for 1,000 simulations.

Before proceeding with the optimization problem, it should be tested whether the GPC works as a good metamodel with respect to the original model. For this purpose, the PDF of the acoustic pressure estimated using the GPC approach is compared with that obtained by the MC simulation, as shown in Figure 7.12. The GPC is obtained from an order $p = 3$ (hence 64 evaluations) through the regression approach. The results show a good match for the shape of the distribution with a slight difference in the mean value of the pressure, which implies that the GPC can be used for the optimization problem.

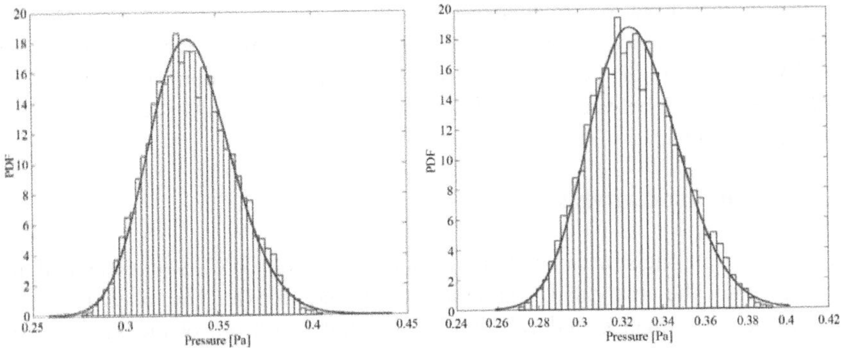

Figure 7.12. *Cavity acoustic pressure distribution (left: MC, right: GPC)*

7.4.3. OSF method coupled with GPC

In order to obtain the best design of the structure, we introduce the RBDO approach combined with the analytical formulation of the coupled cavity example.

The RBDO results are presented in Table 7.6. It is necessary to mention that although the optimal solution satisfies the required pressure limit within the cavity, the reliability level found is lower than the target reliability level $\beta_t = 3.35$ corresponding to a failure probability $P_f = 0.04\%$.

	Parameters	Initial values	Design point	Optimal point
Design variables	h (mm)	4.9	2	2
	F (N)	1.49	1.16	1.06
Objective function	W (kg)	1.98	0.81	0.81
Stress optimization	P_{max} (Pa)	56.57	45.02	40.82
Reliability index	β	–	–	1.028

Table 7.6. *DDO result*

It is for this reason that this method, OSF coupled with GPC, is used to ensure the robustness and reliability of the system. The robustness is guaranteed by the estimation of the mean and variance of the response. Table 7.7 shows the results of this approach.

	Parameters	Initial values	Design point	S_f	Optimal point
Design variables	h (mm)	4.9	2	0.67	3
	F (N)	1.49	1.164	1.01	1.156
Objective function	W (kg)	1.98	0.81	–	1.22
Stress optimization	P_{max} (Pa)	56.57	45.02	–	36.91
Reliability index	β	–	–	–	3.35

Table 7.7. *Results of the OSF coupled with the GPC*

It is remarkable that with the OSF approach, the computational cost is very low as compared to other RBDO methods. Table 7.8 presents a comparison for the optimal point between DDO and OSF using the GPC.

	DDO	OSF
Mass (kg)	0.81	1.22
Maximum pressure (Pa)	40.82	36.91
Level of reliability	1.028	3.35
CPU (s)	3,593.19	2,158.79

Table 7.8. *Comparison of the optimal solution between DDO and OSF*

From this table, we can see that:

– for the deterministic optimization using GPC, the maximum acoustic pressure in the frequency band $[0-500]$Hz is 40.82 Pa, which satisfies the mechanical stress limitation (less than $\frac{50}{S_f} = 41.67$ Pa), with the mass of the structure weighing 0.81 kg, but the reliability level is still very low ($\beta = 1{,}028$) and as such does not guarantee the target reliability and thus has a higher risk of failure;

– in the reliability optimization (OSF using GPC), the maximum acoustic pressure in the frequency band $[0-500]$Hz is 36.91 Pa, which satisfies the mechanical stress limitation (less than 50 Pa), while the mass of the structure is 1.22 kg.

It is necessary to make sure that the GPC model is also well fitted to MC after the optimization procedure. For this reason, the PDF for the acoustic pressure of the optimal solution is shown in Figure 7.13. Table 7.9 shows a comparison of the relative error between GPC and MC for the initial point as well as the optimal solution.

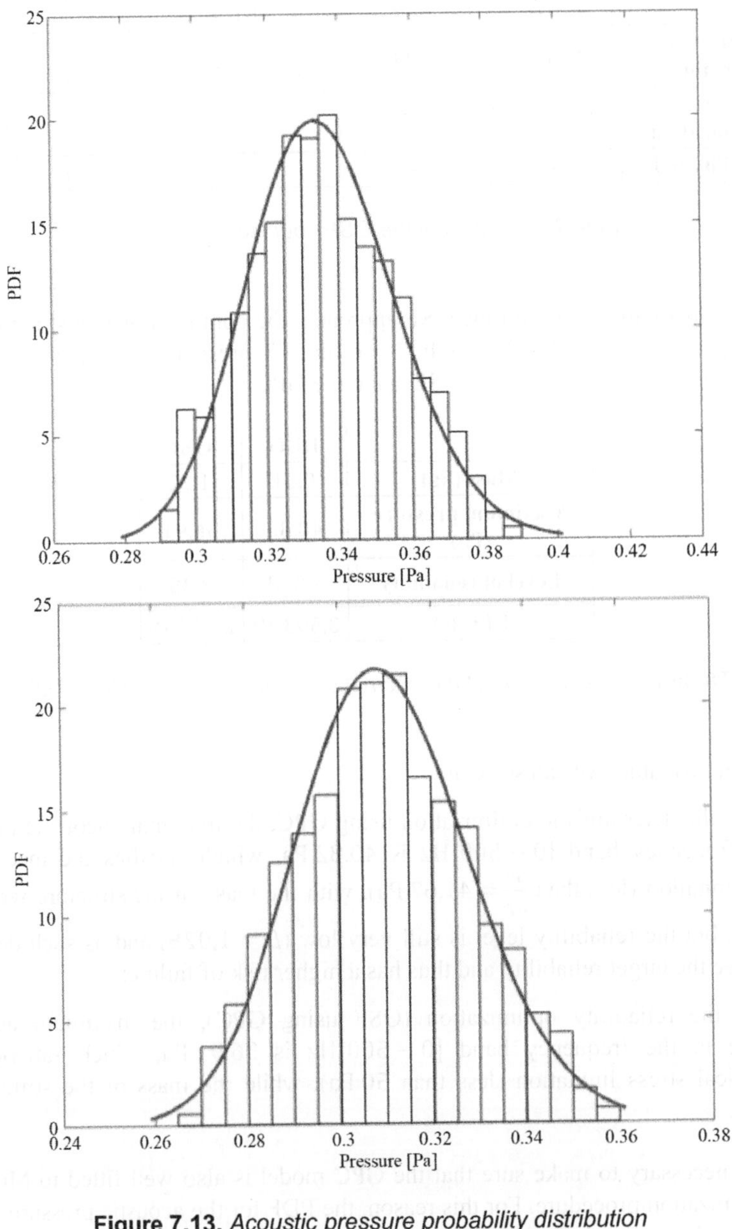

Figure 7.13. *Acoustic pressure probability distribution for the optimal solution (left: MC, right: GPC)*

	Initial solution	Optimal solution
Maximum relative error	6.24%	7.86%

Table 7.9. *Comparison of the relative error between GPC and MC*

7.5. Conclusion

The objective of reliability-based optimization is to define the best compromise between cost reduction and reliability assurance. This chapter is devoted to the study of RBDO for vibro-acoustic problems.

In the first part, the robust approach based on the GPC is employed, for the propagation of uncertainties in the model, in order to circumvent the major drawback of the MC referential method, which is prohibitive in terms of the computation costs for industrial systems.

Next, the different RBDO methods are recalled. We show at the beginning that the implementation of the DDO method is easy. Therefore, DDO can guarantee an optimal design, but cannot control the level of reliability. In addition, RBDO methods such as the HM and the OSF method have been applied. Compared with the DDO method, the RBDO model meets a target reliability level.

In a second part, we considered two applications that take into account the presence of uncertainties. A probabilistic analysis is provided during the optimization procedure. In the end, the chosen approach consists of coupling the OSF method with the GPC method to ensure the robustness of the system during optimization.

7.6. References

[ABI 20] ABID F., DAMMAK K., EL HAMI A., et al., "Surrogate models for uncertainty analysis of micro-actuator", *Microsystem Technologies*, vol. 26, pp. 2589–2600, 2020.

[ABO 11a] ABO AL-KHEER A., EID M., AOUES Y., et al., "Theoretical analysis of the spatial variability in tillage forces for fatigue analysis of tillage machines", *Journal of Terramechanics*, vol. 48, no. 4, pp. 285–295, 2011.

[ABO 11b] ABO AL-KHEER A., KHARMANDA M.G., EL HAMI A., et al., "Estimating the variability of tillage forces on a chisel plough shank by modeling the variability of tillage system parameters", *Computers and Electronics in Agriculture*, vol. 78, no. 1, pp. 61–70, 2011.

[ABO 11c] ABO AL-KHEER A., EL HAMI A., KHARMANDA M.G., et al., "Reliability-based design for soil tillage machines", *Journal of Terramechanics*, vol. 48, no. 1, pp. 57–64, 2011.

[BAR 04] BARAKAT S., BANI-HANI K., TAHA M.Q., "Multi-objective reliability-based optimization of prestressed concrete beams", *Structural Safety*, vol. 26, no. 3, pp. 311–342, 2004.

[BEN 13] BEN ABDESSALEM A., PAGNACCO E., EL HAMI A., "Increasing the stability of T-shape tube hydroforming process under stochastic framework", *International Journal of Advanced Manufacturing Technology*, vol. 69, nos 5–8, pp. 1343–1357, 2013.

[BEN 15] BEN ABDESSALEM A., EL HAMI A., "A probabilistic approach for optimising hydroformed structures using local surrogate models to control failures", *International Journal of Mechanical Sciences*, vols 96–97, pp. 143–162, 2015.

[BOU 96] BOUYEKHF R., EL MOUDNI A., EL HAMI A., et al., "Reduced order modelling of two-time-scale discrete non-linear systems", *Journal of the Franklin Institute*, vol. 333, no. 4, pp. 499–512, 1996.

[BOU 01] BOUYEKHF R., EL HAMI A., EL MOUDNI A., "Optimal control of a particular class of singularly perturbed nonlinear discrete-time systems", *IEEE Transactions on Automatic Control*, vol. 46, no. 7, pp. 1097–1101, 2001.

[BOU 09] BOUAZIZI M.L., GHANMI S., BOUHADDI N., "Multi-objective optimization in dynamics of the structures with nonlinear behavior: Contributions of the metamodels", *Finite Elements in Analysis and Design*, vol. 45, no. 10, pp. 612–623, 2009.

[CHE 08] CHEROUAT A., RADI B., EL HAMI A., "The frictional contact of the composite fabric's shaping", *Acta Mechanica*, vol. 199, nos 1–4, pp. 29–41, 2008.

[CHO 12] CHO J.R., LEE J.H., JEONG K.M., et al., "Optimum design of run-flat tire insert rubber by genetic algorithm", *Finite Elements in Analysis and Design*, vol. 52, pp. 60–70, 2012.

[DAM 17] DAMMAK K., EL HAMI A., KOUBAA S., et al., "Reliability based design optimization of coupled acoustic-structure system using generalized polynomial chaos", *International Journal of Mechanical Sciences*, vol. 134, pp. 75–84, 2017.

[DAM 19a] DAMMAK K., EL HAMI A., "Multi-objective reliability-based design optimization of coupled acoustic-structural system", *Engineering Structures*, vol. 197, pp. 109389, 2019.

[DAM 19b] DAMMAK K., KOUBAA S., EL HAMI A., et al., "Numerical modelling of vibro-acoustic problem in presence of uncertainty: Application to a vehicle cabin", *Applied Acoustics*, vol. 144, pp. 113–123, 2019.

[DAM 20a] DAMMAK K., EL HAMI A., "Multi-objective reliability-based design optimization using Kriging surrogate model for cementless hip prosthesis", *Computer Methods in Biomechanics and Biomedical Engineering*, vol. 23, no. 12, pp. 854–865, 2020.

[DAM 20b] DAMMAK K., YAICH A., EL HAMI A., et al., "An efficient optimization based on the robust hybrid method for the coupled acoustic–structural system", *Mechanics of Advanced Materials and Structures*, vol. 27, no. 21, pp. 1816–1826, 2020.

[DAM 21] DAMMAK K., EL HAMI A., "Thermal reliability-based design optimization using Kriging model of PCM based pin fin heat sink", *International Journal of Heat and Mass Transfer*, vol. 166, pp. 120745, 2021.

[DAS 11] DASKILEWICZ M.J., GERMAN B.J., TAKAHASHI T.T., et al., "Effects of disciplinary uncertainty on multi-objective optimization in aircraft conceptual design", *Structural and Multidisciplinary Optimization*, vol. 44, no. 6, pp. 831–846, 2011.

[DEB 02] DEB K., PRATAP A., AGARWAL S., et al., "A fast and elitist multiobjective genetic algorithm: NSGA-II", *IEEE Transactions on Evolutionary Computation*, vol. 6, no. 2, pp. 182–195, 2002.

[DEB 07] DEB K., PADMANABHAN D., GUPTA S., et al., "Reliability-based multi-objective optimization using evolutionary algorithms", *International Conference on Evolutionary Multi-Criterion Optimization*, Springer, Berlin, Heidelberg, pp. 66–80, 2007.

[DEB 11] DEB K., "Multi-objective optimization using evolutionary algorithms: An introduction", *Multi-objective Evolutionary Optimization for Product Design and Manufacturing*, Springer, London, pp. 3–34, 2011.

[DEB 20] DEBICH B., EL HAMI A., YAICH A., et al., "An efficient reliability-based design optimization study for PCM-based heat-sink used for cooling electronic devices", *Mechanics of Advanced Materials and Structures*, vol. 29, no. 12, pp. 1661–1673, https://doi.org/10.1080/15376494.2020.1836291, 2020.

[DEB 21] DEBICH B., Modélisation, simulation, et fiabilité des matériaux changement de phase pour l'automobile et l'aéronautique. Thesis, INSA Rouen Normandie, 2021.

[DEY 15] DEY S., MUKHOPADHYAY T., KHODAPARAST H.H., et al., "Rotational and ply-level uncertainty in response of composite shallow conical shells", *Composite Structures*, vol. 131, pp. 594–605, 2015.

[DEY 17] DEY S., MUKHOPADHYAY T., ADHIKARI S., "Metamodel based high-fidelity stochastic analysis of composite laminates: A concise review with critical comparative assessment", *Composite Structures*, vol. 171, pp. 227–250, 2017.

[ELH 93] EL HAMI A., LALLEMENT G., MINOTTI P., et al., "Methods that combine finite group theory with component mode synthesis in the analysis of repetitive structures", *Computers and Structures*, vol. 48, no. 6, pp. 975–982, 1993.

[ELH 96] EL HAMI A., RADI B., "Some decomposition methods in the analysis of repetitive structures", *Computers and Structures*, vol. 58, no. 5, pp. 973–980, 1996.

[ELH 09] EL HAMI A., RADI B., CHEROUAT A., "Treatment of the composite fabric's shaping using a Lagrangian formulation", *Mathematical and Computer Modelling*, vol. 49, nos 7–8, pp. 1337–1349, 2009.

[ELH 11] EL HAMI A., RADI B., "Comparison study of different Reliability-Based Design Optimization approaches", *Advanced Materials Research*, vol. 274, pp. 113–121, 2011.

[ELH 13a] EL HAMI A., RADI B., *Incertitude et optimisation et fiabilité des structures*, Hermes-Lavoisier, Paris, 2013.

[ELH 13b] EL HAMI A., RADI B., *Uncertainty and Optimization in Structural Mechanics*, ISTE Ltd, London and John Wiley & Sons, New York, 2013.

[FAN 15] FANG J., GAO Y., SUN G. et al., "Multi-objective robust design optimization of fatigue life for a truck cab", *Reliability Engineering and System Safety*, vol. 135, pp. 1–8, 2015.

[FOR 08] FORRESTER A., SOBESTER A., KEANE A., *Engineering Design via Surrogate Modelling: A Practical Guide*, John Wiley & Sons, New York, 2008.

[FOR 09] FORRESTER A., KEANE, A., "Recent advances in surrogate-based optimization", *Progress in Aerospace Sciences*, vol. 45, nos 1–3, pp. 50–79, 2009.

[GOE 07] GOEL T., VAIDYANATHAN R., HAFTKA R.T., et al., "Response surface approximation of Pareto optimal front in multi-objective optimization", *Computer Methods in Applied Mechanics and Engineering*, vol. 196, nos 4–6, pp. 879–893, 2007.

[GUE 15a] GUERINE A., EL HAMI A., WALHA L., et al. "Approach for the dynamic analysis of one stage gear system with uncertain parameters", *Mechanism and Machine Theory*, vol. 92, pp. 113–126, 2015.

[GUE 15b] GUERINE A., EL HAMI A., FAKHFAKH T., et al., "A polynomial chaos method to the analysis of the dynamic behavior of spur gear system", *Structural Engineering and Mechanics*, vol. 53, no. 4, pp. 819–831, 2015.

[GUE 16a] GUERINE A., EL HAMI A., WALHA L., et al., "A polynomial chaos method for the analysis of the dynamic behavior of uncertain gear friction system", *European Journal of Mechanics A/Solids*, vol. 59, pp. 76–84, 2016.

[GUE 16b] GUERINE A., EL HAMI A., WALHA L., et al., "Dynamic response of a Spur gear system with uncertain friction coefficient", *Advances in Engineering Software*, vol. 120, pp. 45–54, 2016.

[GUE 17] GUERINE A., EL HAMI A., WALHA L., et al., "Dynamic response of wind turbine gear system with uncertain-but-bounded parameters using interval analysis method", *Renewable Energy*, vol. 113, pp. 679–687, 2017.

[HUA 16] HUANG C., RADI B., EL HAMI A., "Uncertainty analysis of deep drawing using surrogate model based probabilistic method", *Manufacturing Technology*, vol. 86, nos 9–12, pp. 3229–3240, 2016.

[JAN 17] JANNOUN M., AOUES Y., PAGNACCO E., et al., "Probabilistic fatigue damage estimation of embedded electronic solder joints under random vibration", *Microelectronics Reliability*, vol. 78, pp. 249–257, 2017.

[KAM 20] KAMEL A., DAMMAK K., EL HAMI A., et al., "A modified hybrid method for a reliability-based design optimization applied to an offshore wind turbine", *Mechanics of Advanced Materials and Structures*, vol. 29, no. 9, pp. 1229–1242, available at: https://doi.org/10.1080/15376494.2020.1811927, 2020.

[KHA 08] KHARMANDA G., MOHSINE A., MAKLOUFI A., et al., "Recent methodologies for reliability-based design optimization", *International Journal for Simulation and Multidisciplinary Design Optimization*, vol. 2, no. 1, pp. 11–23, 2008.

[KHA 09] KHARMANDA G., SHARABATEY S., IBRAHIM H., et al., "Reliability-based design optimization using semi-numerical strategies for structural engineering applications", *International Journal of CAD/CAM*, vol. 9, no. 1, pp. 1–16, 2009.

[KUR 02] KURTARAN H., ESKANDARIAN A., MARZOUGUI D., et al., "Crashworthiness design optimization using successive response surface approximations", *Computational Mechanics*, vol. 29, no. 4, pp. 409–421, 2002.

[LAU 13] LAURENT L., BOUCARD P.A., SOULIER B., "Generation of a cokriging metamodel using a multiparametric strategy", *Computational Mechanics*, vol. 51, no. 2, pp. 151–169, 2013.

[LED 17a] LEDOUX M., EL HAMI A., *Fluid Mechanics: Analytical Methods*, ISTE Ltd, London and John Wiley & Sons, New York, 2017.

[LED 17b] LEDOUX M., EL HAMI A., *Compressible Flow, Propulsion and Digital Approaches in Fluid Dynamics*, ISTE Ltd, London and Wiley & Sons, New York, 2017.

[LED 21a] LEDOUX M., EL HAMI A., *Heat Transfer 1. Conduction*, ISTE Ltd, London and John Wiley & Sons, New York, 2021.

[LED 21b] LEDOUX M., EL HAMI A., *Heat Transfer 2. Radiative Transfer*, ISTE Ltd, London and John Wiley & Sons, New York, 2021.

[LI 19] LI Z., SU X., TAN J., et al., "Multi-objective optimization of the layout of damping material for reducing the structure-borne noise of thin-walled structures", *Thin-Walled Structures*, vol. 140, pp. 331–341, 2019.

[LIA 08] LIAO X., LI Q., YANG X., et al., "Multiobjective optimization for crash safety design of vehicles using stepwise regression model", *Structural and Multidisciplinary Optimization*, vol. 35, no. 6, pp. 561–569, 2008.

[MAB 17] MABROUK I.B., EL HAMI A., WALHA L., et al., "Dynamic vibrations in wind energy systems: Application to vertical axis wind turbine", *Mechanical Systems and Signal Processing*, vol. 85, pp. 396–414, 2017.

[MAC 06] MACKERLE J., "Finite element modeling and simulations in orthopedics: A bibliography 1998–2005", *Computer Methods in Biomechanics and Biomedical Engineering*, vol. 9, no. 3, pp. 149–199, 2006.

[MAK 16] MAKHLOUFI A., AOUES Y., EL HAMI A., "Reliability based design optimization of wire bonding in power microelectronic devices", *Microsystem Technologies*, vol. 22, no. 12, pp. 2737–2748, 2016.

[MCK 00] MCKAY M.D., BECKMAN R.J., CONOVER W.J., "A comparison of three methods for selecting values of input variables in the analysis of output from a computer code", *Technometrics*, vol. 42, no. 1, pp. 55–61, 2000.

[MEC 01] MECKESHEIMER M., BARTON R.R., SIMPSON T.W., et al., "Computationally inexpensive metamodel assessment strategies", *International Design Engineering Technical Conferences and Computers and Information in Engineering Conference*, vol. 80227, American Society of Mechanical Engineers, pp. 191–201, 2001.

[MOH 10] MOHSINE A., EL HAMI A., "A robust study of reliability-based optimization methods under eigen-frequency", *Computer Methods in Applied Mechanics and Engineering*, vol. 199, nos 17–20, pp. 1006–1018, 2010.

[MOR 02] MORO T., EL HAMI A., EL MOUDNI A., "Reliability analysis of a mechanical contact between deformable solids", *Probabilistic Engineering Mechanics*, vol. 17, no. 3, pp. 227–232, 2002.

[MUK 17] MUKHOPADHYAY T., CHAKRABORTY S., DEY S., et al., "A critical assessment of Kriging model variants for high-fidelity uncertainty quantification in dynamics of composite shells", *Archives of Computational Methods in Engineering*, vol. 24, no. 3, pp. 495–518, 2017.

[MYE 16] MYERS R.H., MONTGOMERY D.C., ANDERSON-COOK C.M., *Response Surface Methodology: Process and Product Optimization Using Designed Experiments*, John Wiley & Sons, New York, 2016.

[RAD 07] RADI B., EL HAMI A., "Reliability analysis of the metal forming process", *Mathematical and Computer Modelling*, vol. 45, nos 3–4, pp. 431–439, 2007.

[SAN 03] SANTNER T.J., WILLIAMS B.J., NOTZ W.I, et al., *The Design and Analysis of Computer Experiments*, Springer, New York, 2003.

[SAR 11] SARSRI D., AZRAR L., JEBBOURI A., et al., "Component mode synthesis and polynomial chaos expansions for stochastic frequency functions of large linear FE models", *Computers and Structures*, vol. 89, nos 3–4, pp. 346–356, 2011.

[SIM 01] SIMPSON T.W., POPLINSKI J.D., KOCH P.N., et al., "Metamodels for computer-based engineering design: Survey and recommendations", *Engineering with Computers*, vol. 17, no. 2, pp. 129–150, 2001.

[WAN 17] WANG Y., QIN X., HUANG S., et al., "Structural-borne acoustics analysis and multi-objective optimization by using panel acoustic participation and response surface methodology", *Applied Acoustics*, vol. 116, pp. 139–151, 2017.

[YAI 18] YAICH A., KHARMANDA G., EL HAMI A., et al., "Reliability based design optimization for multiaxial fatigue damage analysis using robust hybrid method", *Journal of Mechanics*, vol. 34, no. 5, pp. 551–566, 2018.

[ZHA 18] ZHANG J., XIAO M., GAO L., et al., "A novel projection outline based active learning method and its combination with Kriging metamodel for hybrid reliability analysis with random and interval variables", *Computer Methods in Applied Mechanics and Engineering*, vol. 341, pp. 32–52, 2018.

List of Authors

Fatma ABID
INSA-Rouen-Normandie
France

Khalil DAMMAK
INSA-Rouen-Normandie
France

Franck DAVENEL
DGA MI
Bruz
France

Ghislaine DELAPAYRE
TechnicAtome
Aix-en-Provence
France

David DELAUX
Valeo
La Verrière
France

Benjamin ECHARD
SECTOR
Lyon
France

Abdelkhalak EL HAMI
INSA-Rouen-Normandie
France

Philippe EUDELINE
Normandie AeroEspace (NAE)
France

Henri GRZESKOWIAK
HGC
Versailles
France

Mohamed HADDAR
ENI-Sfax
Tunisia

Hamid HAMDANI
INSA-Rouen-Normandie
France

Tarek MERZOUKI
UVSQ
France

Lambert PIERRAT
LJC
Gières
France

Bouchaib RADI
FST-Settat
Morocco

Julien RULLIER
SERES
Lyon
France

Hassen TRABELSI
ENI-Sfax
Tunisia

Lassaad WALHA
ENI-Sfax
Tunisia

Index

A, B, C

D, E, F

G, H, I

J, K, L

M, N, O

P, R, S

T, U, V

P, R, S

T, U, V

Summary of Volume 2

Chapter 2. Fatigue Damage Analysis and Reliability Optimization of Structures Subjected to Random Vibrations
Ahmed YAICH and Abdelkhalak EL HAMI

Chapter 3. Accelerated Testing
Henri GRZESKOWIAK, David DELAUX and Abdelkhalak EL HAMI

Chapter 6. Improving Accelerated Reliability Testing by Using Optimized Signals
Jonathan MARTINO

List of Authors

Index

Summary of Volume 3

Other titles from

iSTE

in

Mechanical Engineering and Solid Mechanics

2022

BAYLE Franck
Product Maturity 1: Theoretical Principles and Industrial Applications
(Reliability of Multiphysical Systems Set – Volume 12)
Product Maturity 2: Principles and Illustrations
(Reliability of Multiphysical Systems Set – Volume 13)

EL HAMI Abdelkhalak, DELAUX David, GRZESKOWIAK Henri
Reliability and Physics-of-Healthy in Mechatronics
(Reliability of Multiphysical Systems Set – Volume 15)

LANNOY André
Reliability of Nuclear Power Plants: Methods, Data and Applications
(Reliability of Multiphysical Systems Set – Volume 14)

LEDOUX Michel, EL HAMI Abdelkhalak
Heat Transfer 3: Convection, Fundamentals and Monophasic Flows
(Mathematical and Mechanical Engineering Set – Volume 11)
Heat Transfer 4: Convection, Two-Phase Flows and Special Problems
(Mathematical and Mechanical Engineering Set – Volume 12)

PLANCHETTE Guy
Cindynics, The Science of Danger: A Wake-up Call
(Reliability of Multiphysical Systems Set – Volume 11)

2021

CHALLAMEL Noël, KAPLUNOV Julius, TAKEWAKI Izuru
Modern Trends in Structural and Solid Mechanics 1: Static and Stability
Modern Trends in Structural and Solid Mechanics 2: Vibrations
Modern Trends in Structural and Solid Mechanics 3: Non-deterministic Mechanics

DAHOO Pierre Richard, POUGNET Philippe, EL HAMI Abdelkhalak
Applications and Metrology at Nanometer Scale 1: Smart Materials, Electromagnetic Waves and Uncertainties
(Reliability of Multiphysical Systems Set – Volume 9)
Applications and Metrology at Nanometer Scale 2: Measurement Systems, Quantum Engineering and RBDO Method
(Reliability of Multiphysical Systems Set – Volume 10)

LEDOUX Michel, EL HAMI Abdelkhalak
Heat Transfer 1: Conduction
(Mathematical and Mechanical Engineering Set – Volume 9)
Heat Transfer 2: Radiative Transfer
(Mathematical and Mechanical Engineering Set – Volume 10)

2020

SALENÇON Jean
Elastoplastic Modeling

2019

BAYLE Franck
Reliability of Maintained Systems Subjected to Wear Failure Mechanisms: Theory and Applications
(Reliability of Multiphysical Systems Set – Volume 8)

BEN KAHLA Rabeb, BARKAOUI Abdelwahed, MERZOUKI Tarek
Finite Element Method and Medical Imaging Techniques in Bone Biomechanics
(Mathematical and Mechanical Engineering Set – Volume 8)

IONESCU Ioan R., QUEYREAU Sylvain, PICU Catalin R., SALMAN Oguz Umut
Mechanics and Physics of Solids at Micro- and Nano-Scales

LE VAN Anh, BOUZIDI Rabah
Lagrangian Mechanics: An Advanced Analytical Approach

MICHELITSCH Thomas, PÉREZ RIASCOS Alejandro, COLLET Bernard, NOWAKOWSKI Andrzej, NICOLLEAU Franck
Fractional Dynamics on Networks and Lattices

SALENÇON Jean
Viscoelastic Modeling for Structural Analysis

VÉNIZÉLOS Georges, EL HAMI Abdelkhalak
Movement Equations 5: Dynamics of a Set of Solids
(Non-deformable Solid Mechanics Set – Volume 5)

2018

BOREL Michel, VÉNIZÉLOS Georges
Movement Equations 4: Equilibriums and Small Movements
(Non-deformable Solid Mechanics Set – Volume 4)

FROSSARD Etienne
Granular Geomaterials Dissipative Mechanics: Theory and Applications in Civil Engineering

RADI Bouchaib, EL HAMI Abdelkhalak
Advanced Numerical Methods with Matlab® 1: Function Approximation and System Resolution
(Mathematical and Mechanical Engineering Set – Volume 6)
Advanced Numerical Methods with Matlab® 2: Resolution of Nonlinear, Differential and Partial Differential Equations
(Mathematical and Mechanical Engineering Set – Volume 7)

SALENÇON Jean
Virtual Work Approach to Mechanical Modeling

2017

BOREL Michel, VÉNIZÉLOS Georges
Movement Equations 2: Mathematical and Methodological Supplements
(Non-deformable Solid Mechanics Set – Volume 2)
Movement Equations 3: Dynamics and Fundamental Principle
(Non-deformable Solid Mechanics Set – Volume 3)

BOUVET Christophe
Mechanics of Aeronautical Solids, Materials and Structures
Mechanics of Aeronautical Composite Materials

BRANCHERIE Delphine, FEISSEL Pierre, BOUVIER Salima, IBRAHIMBEGOVIC Adnan
From Microstructure Investigations to Multiscale Modeling: Bridging the Gap

CHEBEL-MORELLO Brigitte, NICOD Jean-Marc, VARNIER Christophe
From Prognostics and Health Systems Management to Predictive Maintenance 2: Knowledge, Traceability and Decision
(Reliability of Multiphysical Systems Set – Volume 7)

EL HAMI Abdelkhalak, RADI Bouchaib
Dynamics of Large Structures and Inverse Problems
(Mathematical and Mechanical Engineering Set – Volume 5)
Fluid-Structure Interactions and Uncertainties: Ansys and Fluent Tools
(Reliability of Multiphysical Systems Set – Volume 6)

KHARMANDA Ghias, EL HAMI Abdelkhalak
Biomechanics: Optimization, Uncertainties and Reliability
(Reliability of Multiphysical Systems Set – Volume 5)

LEDOUX Michel, EL HAMI Abdelkhalak
Compressible Flow Propulsion and Digital Approaches in Fluid Mechanics
(Mathematical and Mechanical Engineering Set – Volume 4)
Fluid Mechanics: Analytical Methods
(Mathematical and Mechanical Engineering Set – Volume 3)

MORI Yvon
Mechanical Vibrations: Applications to Equipment

2016

BOREL Michel, VÉNIZÉLOS Georges
Movement Equations 1: Location, Kinematics and Kinetics
(Non-deformable Solid Mechanics Set – Volume 1)

BOYARD Nicolas
Heat Transfer in Polymer Composite Materials

CARDON Alain, ITMI Mhamed
New Autonomous Systems
(Reliability of Multiphysical Systems Set – Volume 1)

DAHOO Pierre Richard, POUGNET Philippe, EL HAMI Abdelkhalak
Nanometer-scale Defect Detection Using Polarized Light
(Reliability of Multiphysical Systems Set – Volume 2)

DE SAXCÉ Géry, VALLÉE Claude
Galilean Mechanics and Thermodynamics of Continua

DORMIEUX Luc, KONDO Djimédo
Micromechanics of Fracture and Damage
(Micromechanics Set – Volume 1)

EL HAMI Abdelkhalak, RADI Bouchaib
Stochastic Dynamics of Structures
(Mathematical and Mechanical Engineering Set – Volume 2)

GOURIVEAU Rafael, MEDJAHER Kamal, ZERHOUNI Noureddine
From Prognostics and Health Systems Management to Predictive Maintenance 1: Monitoring and Prognostics
(Reliability of Multiphysical Systems Set – Volume 4)

KHARMANDA Ghias, EL HAMI Abdelkhalak
Reliability in Biomechanics
(Reliability of Multiphysical Systems Set –Volume 3)

MOLIMARD Jérôme
Experimental Mechanics of Solids and Structures

RADI Bouchaib, EL HAMI Abdelkhalak
Material Forming Processes: Simulation, Drawing, Hydroforming and Additive Manufacturing
(Mathematical and Mechanical Engineering Set – Volume 1)

2015

KARLIČIĆ Danilo, MURMU Tony, ADHIKARI Sondipon, MCCARTHY Michael
Non-local Structural Mechanics

SAB Karam, LEBÉE Arthur
Homogenization of Heterogeneous Thin and Thick Plates

2014

ATANACKOVIC M. Teodor, PILIPOVIC Stevan, STANKOVIC Bogoljub, ZORICA Dusan
Fractional Calculus with Applications in Mechanics: Vibrations and Diffusion Processes
Fractional Calculus with Applications in Mechanics: Wave Propagation, Impact and Variational Principles

CIBLAC Thierry, MOREL Jean-Claude
Sustainable Masonry: Stability and Behavior of Structures

ILANKO Sinniah, MONTERRUBIO Luis E., MOCHIDA Yusuke
The Rayleigh–Ritz Method for Structural Analysis

LALANNE Christian
Mechanical Vibration and Shock Analysis – 5-volume series – 3rd edition
Sinusoidal Vibration – Volume 1
Mechanical Shock – Volume 2
Random Vibration – Volume 3
Fatigue Damage – Volume 4
Specification Development – Volume 5

LEMAIRE Maurice
Uncertainty and Mechanics

2013

ADHIKARI Sondipon
Structural Dynamic Analysis with Generalized Damping Models: Analysis

ADHIKARI Sondipon
Structural Dynamic Analysis with Generalized Damping Models: Identification

BAILLY Patrice
Materials and Structures under Shock and Impact

BASTIEN Jérôme, BERNARDIN Frédéric, LAMARQUE Claude-Henri
Non-smooth Deterministic or Stochastic Discrete Dynamical Systems: Applications to Models with Friction or Impact

EL HAMI Abdelkhalak, RADI Bouchaib
Uncertainty and Optimization in Structural Mechanics

KIRILLOV Oleg N., PELINOVSKY Dmitry E.
Nonlinear Physical Systems: Spectral Analysis, Stability and Bifurcations

LUONGO Angelo, ZULLI Daniele
Mathematical Models of Beams and Cables

SALENÇON Jean
Yield Design

2012

DAVIM J. Paulo
Mechanical Engineering Education

DUPEUX Michel, BRACCINI Muriel
Mechanics of Solid Interfaces

ELISHAKOFF Isaac *et al.*
Carbon Nanotubes and Nanosensors: Vibration, Buckling and Ballistic Impact

GRÉDIAC Michel, HILD François
Full-Field Measurements and Identification in Solid Mechanics

GROUS Ammar
Fracture Mechanics – 3-volume series
Analysis of Reliability and Quality Control – Volume 1
Applied Reliability – Volume 2
Applied Quality Control – Volume 3

RECHO Naman
Fracture Mechanics and Crack Growth

2011

KRYSINSKI Tomasz, MALBURET François
Mechanical Instability

SOUSTELLE Michel
An Introduction to Chemical Kinetics

2010

BREITKOPF Piotr, FILOMENO COELHO Rajan
Multidisciplinary Design Optimization in Computational Mechanics

DAVIM J. Paulo
Biotribolgy

PAULTRE Patrick
Dynamics of Structures

SOUSTELLE Michel
Handbook of Heterogenous Kinetics

2009

BERLIOZ Alain, TROMPETTE Philippe
Solid Mechanics using the Finite Element Method

LEMAIRE Maurice
Structural Reliability

2007

GIRARD Alain, ROY Nicolas
Structural Dynamics in Industry

GUINEBRETIÈRE René
X-ray Diffraction by Polycrystalline Materials

KRYSINSKI Tomasz, MALBURET François
Mechanical Vibrations

KUNDU Tribikram
Advanced Ultrasonic Methods for Material and Structure Inspection

SIH George C. *et al.*
Particle and Continuum Aspects of Mesomechanics

Printed and bound by CPI Group (UK) Ltd, Croydon, CR0 4YY
16/09/2026
14963209-0001